STUDENT UNIT GUIDE

UNIT

OCR A2 F763

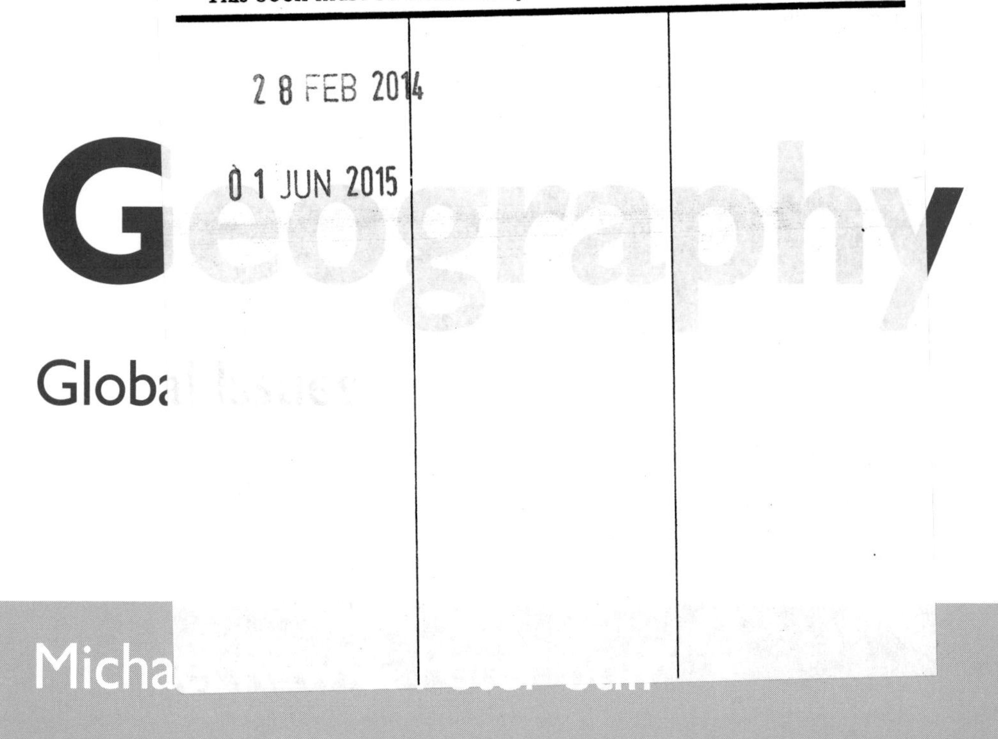

Geography

Globa

Micha

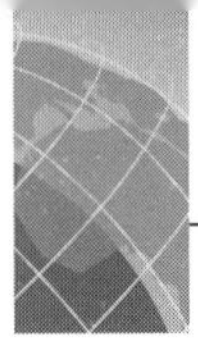

With thanks to Katrina Hillary and Sarah Whitehead, and to Alison for yet more encouragement

Philip Allan Updates, an imprint of Hodder Education, an Hachette UK company, Market Place, Deddington, Oxfordshire OX15 0SE

Orders
Bookpoint Ltd, 130 Milton Park, Abingdon, Oxfordshire, OX14 4SB
tel: 01235 827827
fax: 01235 400401
e-mail: education@bookpoint.co.uk
Lines are open 9.00 a.m.–5.00 p.m., Monday to Saturday, with a 24-hour message answering service. You can also order through the Philip Allan Updates website: www.philipallan.co.uk

ISBN 978-0-340-99086-5
First printed 2010

Impression number 5 4 3
Year 2014 2013 2012

Typeset by Philip Allan Updates

Printed by MPG Books, Bodmin

Contents

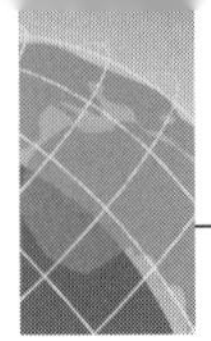

Introduction

About this guide

This guide is designed to help you prepare for OCR A2 Geography **Unit F763 Global Issues**. It is divided into three parts: an introduction; a content guidance section; and a question and answer section.

This **Introduction** explains the assessment structure and outlines the techniques for dealing with both data-response and extended writing questions. The **Content Guidance** section outlines the specification content and the key themes used to formulate examination questions. The **Questions and Answers** section provides 12 specimen questions (six data-response questions and six extended writing (essay) questions) with one student answer for each question, ranging from grade A* to grade E. Student answers are followed by examiner's comments. You are advised to study all of the questions and answers in this guide, and not just concentrate on those that relate to the options you are studying: examiner's comments include general points of advice applicable to all questions in the unit.

Assessment

F763 Global Issues is one of two units that make up the A2 specification. It is worth 120 uniform marks and accounts for 60% of the marks at A2 and 30% of the total specification, AS + A2.

The other A2 unit is F764: Geographical Skills (see Table 1).

Table 1 A2 Geography: scheme of assessment

Unit number and unit name (exam length)	Raw marks	Uniform marks (A2 weighting)
F763: Global Issues (2.5 hours)	90	100 (60)
F764: Geographical Skills (1.5 hours)	60	80 (40)

Unit F763 covers six major topics (options), three on environmental issues and three on economic issues:

Environmental issues
- Earth hazards
- Ecosystems and environments under threat
- Climatic hazards

Economic issues
- Population and resources

- Globalisation
- Development and inequalities

The exam paper is in two parts, Section A and Section B. In **Section A** you have to answer **three** data-response questions. You must answer at least **one** from **three** questions on Environmental issues and at least **one** from **three** questions on Economic issues.

Section B requires you to answer **two** extended writing questions. You must answer **one** from **six** questions on Environmental issues and **one** from **six** questions on Economic issues. The requirement of the question paper means, therefore, that you must study at least three of the six topics (options) specified for Unit F763.

Data-response questions

The data-response questions in Section A are each worth 10 marks. All of the questions have the same wording, asking you to *'Outline a geographical issue indicated and suggest appropriate strategies for its management.'* A wide range of stimulus material is used with these questions, including OS maps, sketch maps, charts, photographs, satellite images and text. Normally one resource per question will be used, though occasionally the examiners might use more than a single resource, such as a pair of photographs. In total, the data-response questions are worth 30 out of the 90 raw marks available for Unit F763. Thus in a 2.5 hour exam, you would expect to spend approximately 20 minutes on each data-response question.

Extended writing (essay) questions

Section B requires you to answer two extended writing (essay) questions, each worth 30 marks. You should allow approximately 45 minutes for each essay, making 1.5 hours in total. These questions are demanding, requiring a structured discussion supported by case studies and examples. You should be able to describe, explain, analyse, apply, and, most importantly, evaluate the issue on which the question is based.

Mark scheme criteria

Examination answers are assessed against three criteria or assessment objectives (AOs). For A2 Geography these are as follows:

1 **AO1 — Demonstrate knowledge and understanding** of the content, concepts and processes.

2 **AO2 — Analyse, interpret and evaluate** geographical information, issues and viewpoints, and apply understanding in unfamiliar contexts.

3 **AO3 — Investigate, conclude and communicate**: selecting and use a variety of methods, skills and techniques (including new technologies) to investigate questions and issues, reach conclusions and communicate findings.

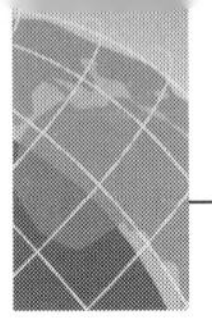

You should study these criteria carefully because they describe how your examination answers will be judged. The section on examination skills below explains how assessment objectives are used in mark schemes to assess your answers. Table 2 shows the weightings given to each AO.

Table 2 Assessment objective weightings in A-level Geography (AS + A2)

	% of A-level Geography			
Unit number and unit name	**AO1**	**AO2**	**AO3**	**Total %**
A2 Unit F763: Global Issues	10	15	5	30
A2 Unit F764: Geographical Skills	5	5	10	20

Examination skills

Success in A2 Geography requires not only sound knowledge and understanding of the specification content, but also effective exam technique. To acquire a solid knowledge base, you should structure your revision around the key ideas and key questions set out in the **Content Guidance** section of this guide. This structure will help focus your learning on the areas most frequently targeted by examiners. To achieve the higher grades in the more demanding questions, you must be able to apply your knowledge and understanding accurately and in unfamiliar contexts. Failure to do this is a common cause of low marks.

Answering data-response questions

The stimulus material provides a prompt to the issue in question. You are not expected to have detailed knowledge about the actual example used in the stimulus material, but you will need to respond to the underlying issue it raises. The management strategies you suggest must be appropriate for the identified issue.

The mark scheme for the data-response questions is levels based. There are three levels of attainment with each level defined by a descriptor (see Table 3). Having read your answer, the examiner will first allocate it to a level, and then decide the precise mark.

Table 3 Generic mark scheme used to assess Section A data-response questions

Level	Marks	Descriptor
3	9–10	Substantial knowledge and authoritative understanding of an appropriate geographical issue. Clear application of relevant knowledge and understanding to the question set including the suggestion of appropriate management strategies. Clear structure and organisation. Communication is clear with maps, diagrams, and statistics, where appropriate. Accurate use of geographical terms.

Level	Marks	Descriptor
2	5–8	Sound knowledge and understanding of an appropriate geographical issue. Sound application of relevant knowledge and understanding to the question set including the suggestion of appropriate management strategies. Sound structure and organisation. Communication is generally effective with maps, diagrams, and statistics, if appropriate. Geographical terms are mainly used effectively.
1	0–4	Poor knowledge and understanding of an appropriate geographical issue or inappropriate issue. Limited application of relevant knowledge and understanding to the question set such as the suggestion of inappropriate management strategies or none suggested. Poor structure and organisation. Much inaccuracy in communication and limited and/or ineffective use of geographical terms.

When answering the data-response questions in Section A, you should follow these guidelines:

- Spend a couple of minutes examining the stimulus material. You need to be sure of an appropriate geographical issue highlighted by the resource before writing your answer.
- Start your answer with a short paragraph that clearly identifies the focus of the question.
- Identify, in a single sentence, the geographical issue that you are discussing.
- Suggest two or three management strategies for dealing with the issue. These must be appropriate to the issue in question.
- Consider the appropriateness of long- and/or short-term management strategies.
- Consider the scale of the management strategy. Some strategies will be implemented by local communities, some by national governments and some by international agreements.
- Watch the clock! Do not exceed the allocation of 20 minutes for each data-response question.

Answering extended writing questions

Answers are assessed against the same three criteria or assessment objectives (AOs) used for data-response questions. Each assessment objective is divided into three attainment levels see (Table 4).

Table 4 Generic mark scheme used to assess Section B essay questions

AO1 Knowledge and understanding

Descriptor	Marks
Level 3 Substantial knowledge and authoritative understanding of the appropriate issue(s) and management strategies.	8–9 marks
Level 2 Sound knowledge and understanding of the appropriate issue(s) and management strategies.	5–7 marks

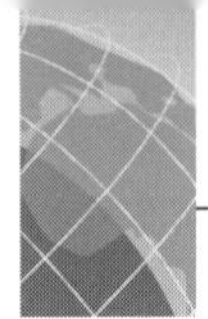

Level 1 Poor knowledge and understanding of the appropriate issue(s) and management strategies.	0–4 marks

AO2 Analysis, interpretation and evaluation

Level 3 Clear analysis of relevant issues and management strategies. Thoughtful and effective evaluation drawing together relevant geographical information, issues and viewpoints.	14–17 marks
Level 2 Sound analysis of relevant issues and management strategies. Some evaluation drawing together relevant geographical information, issues and viewpoints.	8–13 marks
Level 3 Limited analysis of relevant issues and management strategies. No evaluation drawing together geographical information, issues and viewpoints.	0–7 marks

AO3 Investigation, conclusion and communication

Level 3 Clear structure and organisation including introduction and conclusion. Communication is clear with accurate grammar and spelling. Confident and precise use of geographical terms. Focused use of maps, diagrams, statistics, if appropriate.	4 marks
Level 2 Sound structure and organisation including introduction and conclusion. Communication is sound but with some inaccurate grammar and spelling. Use of geographical terms imprecise. Some use of maps, diagrams, statistics, if appropriate.	3 marks
Level 1 Poor structure and organisation not including one or both of introduction and conclusion. Communication is basic with inaccurate grammar and spelling. Little or no use of geographical terms. Ineffective or no use of maps, diagrams, statistics, if appropriate.	1–2 marks

The majority of the marks are awarded under AO2, which underlines the importance of evaluation in your answers. Each question requires a discussion of issues for which there is rarely a clear-cut answer. You are expected to debate the arguments in full, supporting them with examples. General lines of discussion that are often relevant include the importance of scale, and the distinction between places (countries, regions) at different levels of economic and social development.

Just under a third of the marks available are for knowledge and understanding (AO1). It is clear that you must have a secure grasp of geographical patterns and processes as well as facts and figures to support your arguments. Although AO3 is worth only a maximum of 4 marks, your answer will inevitably suffer in AO1 and AO2 if your discussion is poorly organised and poorly expressed. Also be aware of opportunities to include simple sketch maps or sketch diagrams in your answers. Sometimes they are more effective devices for communicating knowledge and understanding than text.

When answering extended writing questions, you should follow these guidelines:

- Spend 2 or 3 minutes reflecting on the question and writing a plan.
- In your plan, outline the general content of each paragraph of the answer and the geographical examples you intend to use to support your discussion.
- Make sure that your answer has an introduction, a main body and a conclusion.
- **Introduction**: this should 'set the scene' for your discussion. Define the key terms such as tropical storm, extreme weather, resource, globalisation or quality of life. Introductions should be brief and not attempt to offer answers to the question.
- **Main body**: this is where you develop the points raised in your introduction. Each point should have its own paragraph in which you debate its role in the context of the issue under discussion. You will need to support each argument with real world examples of direct relevance. There is no value in adopting a 'write all I can remember' approach to a question. Such an approach would be indiscriminate, lack relevance and be heavily penalised. Try to provide a diversity of perspectives — for example, the impacts of similar hazards in contrasting environments or the advantages and disadvantages of foreign direct investment on countries at different ends of the development spectrum.
- **Conclusion**: there should be a brief summary of the arguments you developed in the main body of your answer but not a simple repetition of them. You should offer a reflective view of the question which (a) makes your values and opinion clear, and (b) is wholly consistent with the arguments developed in the main body of your answer.
- Watch the clock! You need to write two, evenly balanced answers to score high marks. It is no good spending an hour on the first answer and leaving just half an hour for the second. However good your first answer, it will not make up for the loss of marks on the second.

Command words and phrases

Command words and phrases in examination questions are crucial because they tell you exactly what you have to do. You must respond precisely to their instructions. Ignore them at your peril — students that do invariably underachieve.

In Section A, all of the questions have the same format: *'Identify a geographical issue and suggest appropriate management strategies.'* '**Identify**' means to determine or decide the geographical issue shown in the resource. '**Suggest**' tells you to describe possible management strategies.

In Section B, the following command phrases are often used:

'Assess the extent to which...', **'Assess the degree to which...'**, **'Evaluate the relative significance of...'**, **'How far do you agree...?'**

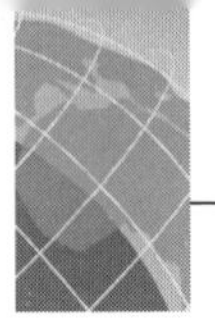

Some questions comprise a brief quote or statement, followed simply by the instruction '**Discuss**'. 'Discuss' (and the command phrases on page 9) ask you to consider the evidence and arguments connected to an issue or problem, make reasoned judgements and present a point of view. Whichever form the command takes in Section B, evaluation must be an important part of your answer.

Unit F763 is synoptic, which means that you should draw widely on knowledge, understanding and skills from other parts of the A-level course and integrate them into your answers. For this reason the extended writing questions at A2 are deliberately broad-based and allow you to make connections with topics and ideas studied at both AS and A2. For example, answers to questions on flood hazards at A2 could refer to AS option studies of River environments and/or Coastal environments. References to relevant fieldwork/research investigations for F764 (Geographical Skills) could also be credited as synoptic.

Case studies

An important feature of the OCR A2 Geography specification is its emphasis on exemplification through detailed case studies. The most effective answers, particularly in the extended writing questions in Section B, will include detailed examples and case studies. In some places the specification makes clear reference to case studies such as 'one named country' or 'two contrasting types of climatic hazards'. Make sure that you read the Content Guidance section carefully so that you are fully aware of the requirements for case studies. Remember this is a geography examination and that it is expected that your answers will demonstrate a solid grounding in real places, real people and real environments.

Content Guidance

This section provides a summary of the key ideas and content detail needed for A2 Geography Unit F763 Global Issues.

The content is divided into six main topics under Environmental issues and Economic issues:

A: Environmental issues

- Earth hazards
- Ecosystems and environments under threat
- Climatic hazards

B: Economic issues

- Population and resources
- Globalisation
- Development and inequalities

When you revise it is important to use a framework that reflects how examiners might test your knowledge and understanding. Therefore, in addition to the key ideas and content detail, this section provides **key questions** and **answers** for each topic.

You should study the questions and answers carefully and organise your revision around them. Focusing on the key questions and adding details of your own to the answers should give you a head start in the final examination.

It is essential that you learn the terminology used in the answers to the key questions, particularly the words in **bold** type. You must be able to apply these terms appropriately in your exam answers.

Environmental issues

Earth hazards

What are the hazards associated with earthquake activity?

Key ideas	Content detail
• Earthquakes are caused by plate tectonics and bring distinctive impacts to an area, which vary from place to place. • Earthquakes have a range of environmental, social and economic impacts on the areas affected, which create a range of human responses. • The impact of an earthquake on an area reflects its level of economic and technological development and its population density. Impacts vary over time from immediate to long term. • There are various ways to manage or reduce the impacts of earthquakes.	• The tectonic processes involved (inter-plate movements, intra-plate movements along fault lines). • Earthquake impacts vary in scale and in their environmental, social and economic effects. Primary hazards include ground shaking, liquefaction, mass movements and fires. Secondary hazards include disease, infrastructure problems, resettlement etc. • The human response to earthquake disasters is both short term (emergency rescue) and long term (planning and management). • The contrast in the impact and reaction to earthquakes between countries at either end of the development continuum and between urban and rural areas. • A comparison of impacts over short and long time periods. • The extent to which earthquakes are predictable. • Management strategies to reduce the impact of earthquake hazards and their effectiveness.

Key questions

What are the causes of earthquakes?

• Earthquakes, caused by sudden movements of the Earth's crust and lithosphere, result in **primary hazards** such as ground shaking, **liquefaction**, landslides and **tsunamis**. Essentially, earthquakes are due to stresses in the crust and lithosphere in the form of compression, tension and shearing. All three processes occur at tectonic plate boundaries. The earthquakes generated there are known as **inter-plate quakes**. Compression occurs at **convergent plate margins** where **subduction** takes place. These compressive forces give rise to low-angled **thrust faults**. Movements along these faults often produce high-magnitude quakes such as the Sumatran quake in 2004. Tensional forces are associated with **divergent plate margins** or **mid-ocean ridges**, where stretching of the crust and lithosphere leads to faulting and **rifting**. Rifting is when blocks of crust slip downwards along a series of parallel faults. Earthquakes at **conservative plate margins** are due to horizontal or shearing movements as two plates slide past each other. This movement produces violent earthquakes (e.g. California).

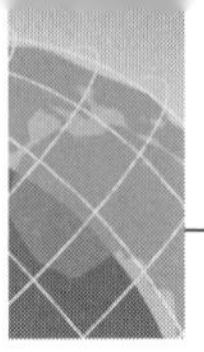

Many large earthquakes occur a long way from tectonic plate margins (e.g. Sichuan quake in 2008). These so-called **intra-plate quakes** also result from compressional and tensional forces that cause rocks to snap and move suddenly along fault lines.

What are the environmental, social and economic impacts of earthquakes?

Earthquakes often trigger mass movements (especially landslides) in mountainous regions such as Kashmir and Sichuan, destroying settlements and farmland, and blocking roads and railways. The Sichuan quake in 2008 led to landslides that dammed rivers and created dozens of temporary barrier lakes. Pollution of the lakes by domestic waste threatened the health of local populations. The Sumatran tsunami caused severe damage to mangroves and coral reefs in Indonesia and Malaysia. Farmland, contaminated by salt water, had to be abandoned and salt water also polluted freshwater aquifers.

Major earthquakes cause large-scale mortality, especially in poorer countries. 87,000 people died in the Kashmir quake (2005) and a similar number in the Sichuan quake. The 2004 tsunami killed an estimated 250,000 people in 11 countries. Large numbers of injuries also result from earthquake events, mainly due to falling masonry and collapsed buildings. The Sichuan quake, for example, made 5.4 million people homeless. Not only are millions of people displaced, but also the threat of disease from contaminated drinking water (e.g. cholera, typhoid) is a real problem. There is often an acute shortage of medicines, doctors and hospitals to treat survivors.

The economic impact of large earthquakes is greatest in the world's richest countries. The Northridge quake in southern California in 1994 killed just 57 people, but in economic terms was the costliest quake in US history (US$20 billion). The quake brought down freeway intersections, ruptured pipelines and damaged bridges. Many steel-framed buildings suffered damage and houses were destroyed by landslides. The Kobe quake in Japan in 1995 destroyed 56,000 buildings and many people were killed when elevated urban freeways collapsed. The port of Kobe, Japan's leading container port, was also badly damaged by the quake.

What are the short-term responses to earthquake disasters?

In the short term, the response to earthquake disasters is to provide emergency rescue and relief to survivors. In major earthquake disasters such as Gujarat in 1999 and Kashmir in 2005, coordinated international emergency relief was needed, with contributions from foreign governments and NGOs (e.g. Red Cross, Green Crescent). The immediate priority is to rescue people trapped beneath collapsed buildings. After a few days, the focus shifts to providing food, water, shelter, heating, medicines and sanitation for survivors. The main threat to survivors is disease caused by inadequate sanitation and contaminated water. However, the nature of emergency relief depends on the location and timing of the disaster. In remote regions like Kashmir, helicopters were essential to reach isolated communities in the mountains. Severe winter weather in the region also meant that heating and shelter were priorities.

What are the long-term responses to earthquake disasters?

The long-term response to earthquake disasters is reconstruction of the devastated areas. Homes, schools, hospitals, businesses, roads, bridges and other essential infrastructure need to be rebuilt, and jobs have to be provided. This is a slow process, which may take decades and cost billions of dollars. For example, the total bill for reconstruction following the Sichuan earthquake will probably exceed US$150 billion. Reconstruction costs following the Kobe earthquake were US$120 billion.

The reconstruction plan for the Sichuan quake shows the enormous scale of the task. It includes 169 new hospitals and nearly 4,500 new primary schools, while 3 million homeless rural families will get new houses, and 860,000 city apartments will be built. In developing countries, reconstruction on this scale requires aid from foreign governments and from multilateral agencies such as the World Bank and the International Monetary Fund.

What factors determine the social and economic impact of earthquake hazards?

Two sets of factors determine the economic and social impact of earthquakes: exposure and vulnerability. **Exposure** combines the magnitude of a quake with the number of people at risk in the affected area. **Vulnerability** describes how resistant a society is to earthquake hazards: it is strongly influenced by levels of **preparedness** and the ability to mitigate earthquake impacts.

Exposure to earthquakes is high in tectonically active regions such as Japan, southern California and eastern China, which are densely populated and where large quakes occur frequently. However, the social and economic impact of earthquakes in regions with similar exposure is highly variable. In wealthy countries like the USA and Japan, death tolls are relatively low, while the economic costs are huge. Thus while the 1994 quake at Northridge in southern California claimed only 57 lives, the economic cost was US$20 billion. In poor countries the number of fatalities and scale of homelessness are usually much higher, but the economic costs are often lower. Thus the Kashmir quake killed an estimated 87,000 people but the economic cost of the quake — US$5 billion — was only a fraction of the cost of the Northridge quake.

Geographical variations in vulnerability explain differences in the impact of earthquakes. Rich countries can afford to build earthquake-resistant societies. Buildings are earthquake-proof and fire-proof; emergency services are highly trained; people are educated in how to respond to earthquake hazards and take responsibility for themselves and their families; and local governments have detailed disaster plans and management strategies. The overall effect of these measures is to lower vulnerability in areas of high exposure and reduce the number of deaths and injuries. Economic losses will also be reduced, but given the scale of investment and the value of the built environment in developed countries, they will generally be much higher than in developing countries.

How can the impact of earthquake hazards be reduced?

Earthquakes cannot be predicted. However, several measures can be taken to reduce their impact. So far, these measures have largely been confined to rich countries. Their success is evident in the relatively small number of deaths and injuries related to earthquake hazards in developed countries in the past 15 years or so.

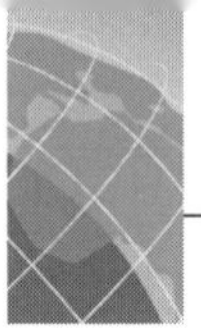

Buildings in earthquake zones can be strengthened by using solid steel frames, reinforced concrete and steel braces. High-rise buildings are particularly vulnerable to ground shaking and liquefaction and may have flexible structures, massive concrete pillars, counterweights and deep foundations to ensure their stability. Buildings can also be made fire-proof. But the construction of earthquake-proof buildings is of little use without mandatory building codes that are properly enforced. The huge loss of life in the Sichuan earthquake was partly due to inadequate enforcement of building codes, shoddy workmanship and poor-quality building materials.

Disaster planning plays a vital part of earthquake hazard mitigation. Tokyo's earthquake disaster plan aims to make the city earthquake-resistant, and involves individuals as well as government. The plan identifies those parts of the city where buildings need upgrading to make them earthquake- and fire-proof. Roads, expressways, bridges and utilities are being strengthened, together with important government offices, police and fire stations. Refuge sites, provided with emergency supplies and shelters have been designated in public parks and other open spaces. The disaster plan also sets out priorities and targets for recovery and reconstruction. Individual households are encouraged to take responsibility for their own safety and well-being and assist the community. To this end, an educational programme raising public awareness of the earthquake hazard has been established for over a decade.

What are the hazards associated with volcanic activity?

Key ideas	Content detail
• Volcanic activity is caused by plate tectonics and brings distinctive impacts to an area, which vary from place to place. • Volcanic eruptions have a range of environmental and social impacts on the areas affected, which create a range of human responses. • The impact of a volcanic eruption on an area reflects its level of economic and technological development and its population density. Impacts vary over time from immediate to long term. • There are various ways to manage or reduce the impacts of volcanoes.	• The tectonic processes involved (convergent plate margins and subduction, divergent margins and sea floor spreading, and hot spots). • Volcanic eruptions vary in scale and in their environmental, social and economic impacts. Primary hazards include lava flows, ashfalls, pyroclastic flows and lahars. Secondary hazards include lack of shelter, food, disease, infrastructure problems, resettlement etc. • The human response to volcanic disasters is both short term (emergency rescue) and long term (planning and management). • The contrast in the impact and reaction to volcanic eruptions between countries at either end of the development continuum and between urban and rural areas. • A comparison of impacts over short and long time periods. • The extent to which volcanic eruptions are predictable. • Management strategies to reduce the impact of volcanic hazards and their effectiveness.

Key questions

What are the causes of volcanic activity?

Volcanic activity is found at three types of location: convergent plate margins, divergent plate margins and hot spots. Subduction occurs at **convergent margins** as one tectonic plate **underthrusts** the other and descends into the upper mantle. At depths of 120 to 180 km the subducted plate begins to melt and the molten rock (magma) rises slowly to the surface. Where it breaks the surface it forms a **volcano** or volcanic **fissure**. The Lesser Antilles chain of volcanic islands has been formed by subduction. In this example, the North American plate has been subducted beneath the Caribbean plate forming volcanoes like Soufrière Hills on Montserrat. Volcanoes associated with subduction zones are **andesitic**, explosive and extremely hazardous.

Volcanic activity also occurs at **divergent margins** along mid-ocean ridges and continental locations like east Africa's Great Rift Valley. Tension stretches the crust and creates a series of parallel faults, which leads to rifting. The subsequent reduction in pressure in the upper mantle allows molten rock to flow to the surface. Volcanic islands such as Iceland and Ascension Island in the Atlantic Ocean have formed in this way. Eruptions at divergent margins are generally non-explosive or **effusive**, and comprise basalt lava.

Similar effusive eruptions take place at hot spots like Hawaii, where rising plumes of hot magma have punched a hole through the crust. Mantle plumes remain more or less stationary over millions of years while the overlying tectonic plates move over them. This explains the formation of the volcanic Hawaiian Islands, and the fact that volcanoes on the oldest islands are now extinct.

What are the environmental, social and economic impacts of volcanic eruptions?

The primary hazards of volcanic eruptions are lava flows, ashfalls, pyroclastic flows, lahars and poisonous gases. These hazards have environmental, social and economic impacts.

- **Lava flows** are streams of molten and semi-molten magma erupted from a volcano. Their impact is usually fairly localised. Although rarely a threat to life, they destroy everything in their path — houses, roads, farmland. Lava flows that are non-viscous and not confined by valleys, can spread out to form vast lava fields. Extensive areas of forest, farmland and even entire villages in Hawaii have been buried by lava in this way.
- Hazards caused by volcanic ash are more widespread. Localised **ashfalls** and accumulations of **scoria** can cause buildings to collapse, while farmland may be completely buried under several metres of ash. Ash from major volcanic eruptions is pumped high into the stratosphere and within a few weeks completely encircles the planet. These tiny ash particles reflect solar radiation and can lower average global **temperatures**. The 1991 eruption of Pinatubo resulted in a cooling of the global climate over the next 2 years.

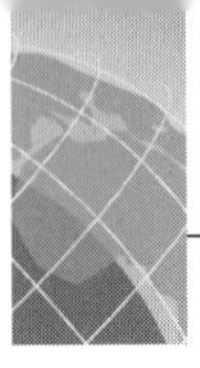

- **Pyroclastic flows** are the most deadly volcanic hazard. They are ground-hugging avalanches of hot ash, pumice, rock fragments and gas. Temperatures within pyroclastic flows may exceed 500°C and they can travel at over 100 km h^{-1}. They destroy everything in their path and can even move across sea and lake surfaces.
- **Lahars** are fast-moving mudflows caused by run-off from heavy rain and/or melting snow transporting large volumes of loose volcanic ash. Lahars follow well-defined paths along river valleys and because of their speed (up to 50 km h^{-1}) are particularly deadly. In 1985, lahars caused by the eruption of Nevado del Ruiz in Colombia devastated the town of Armero and killed 23,000 people.
- Some volcanoes emit large quantities of **toxic gas** (e.g. sulphur dioxide). Vents on the flanks of Nyiragongo release up to 50,000 tonnes of sulphur dioxide a day. These emissions are highly dangerous and threaten human health, livestock, crops and forests. They also acidify rivers and lakes.

What are the short-term responses to volcanic disasters?

Unlike earthquake hazards, most volcanic eruptions can be predicted days or even weeks in advance. The immediate response to an eruption that threatens disaster is to evacuate the population. In 2008, the Chaitén volcano in southern Chile erupted, spreading large quantities of ash over the surrounding area. The risks to the population of the town of Chaitén at the foot of the volcano were such that the entire population (c4,000) were evacuated by sea within 24 hours. Following the eruption of the Soufrière Hills volcano on the Caribbean island of Montserrat in 1995, the southern part of the island was declared an exclusion zone and the population evacuated to safe areas in the north. Emergency shelters were provided and local people were given the option of leaving Montserrat for neighbouring islands.

Following the Nyiragongo eruption in 2002, thousands of refugees were housed in temporary camps. Poor sanitation and overcrowding polluted drinking water from Lake Kivu and threatened a cholera outbreak. Swift action taken by NGOs to provide chlorinated drinking water contained the disease. Aid agencies also provided immunisation against measles, meningitis and polio, and distributed emergency food rations.

What are the long-term responses to volcanic disasters?

Following the emergency response to a volcanic disaster, attention and resources focus on normalising economic and social systems by restoring infrastructure, services and employment. The volcanic eruptions that devastated Montserrat between 1995 and 1997 were followed by a period of reconstruction. Emergency shelters were replaced by low-cost permanent housing and many islanders were resettled. The construction of a new airport helped to rebuild the island's tourism industry. In future, a new observatory staffed by foreign scientists will monitor the Soufrière Hills volcano. Several NGOs have been involved in the long-term reconstruction of Montserrat's economy and infrastructure, and the UK and EU have invested around £200 million on regeneration projects.

What factors determine the social and economic impact of volcanic hazards?

Compared to other natural hazards, the economic and social impact of volcanic eruptions varies with the type of hazard. For example, lava flows mainly affect small areas on the flanks of volcanoes. Ashfalls are more widespread and lahars can devastate communities 100 km or more from the site of eruption. However, the impact of lahars is usually confined to river valleys.

The overall impact of volcanic eruptions depends on two factors: exposure and vulnerability. In connection with volcanic hazards, exposure refers to the scale and violence of an eruption and to the density of population living near the volcano. Andesitic volcanoes like Mount St Helens are **explosive**. These eruptions present a much greater threat than **effusive** eruptions from volcanoes like Kilauea (Hawaii). When Mount Pelée, an andesitic volcano on the island of Martinique, erupted in 1902, pyroclastic flows killed 30,000 in the capital, Saint-Pierre. In contrast, the explosive eruption of Mount St Helens in Washington state in 1980 killed just 57 people. The difference is due to population distribution: the region around Mount St Helens is mainly forested and sparsely populated, so hazard exposure was much lower than near Mount Pelée.

The impact of volcanic hazards is also due to vulnerability. Impacts are mitigated in rich countries such as Italy, Japan and Iceland, where volcanoes can be monitored, evacuation organised and resources for emergency relief and reconstruction programmes mobilised. This level of preparedness is notably absent in most developing countries. As a result, deaths and injuries from lahars and pyroclastic flows are much higher than in developed countries. In the post-disaster period there are also greater threats from disease, polluted water supplies and inadequate medical care.

How can the impact of volcanic hazards be reduced?

Observation of volcanoes and monitoring their behaviour can give early warning of impending eruptions. Seismic activity increases before an eruption as magma forces its way to the surface and fractures rocks inside the volcano. Seismic waves are recorded by networks of seismometers on the volcano. Gravity increases as the **magma chamber** fills and this change is monitored. Gases emitted by **fumaroles** are also sampled. Increased levels of sulphur dioxide and hydrogen chloride often signal an imminent eruption. Finally ground deformation (**inflation**) is measured. As magma accumulates below the surface, the ground surface may rise by 3 or 4 m before an eruption.

Diverting lava flows has, on occasion, been successful. Small lava flows on the flanks of Mount Etna in Sicily have been diverted away from centres of population. At Heimaey in Iceland in 1973, a lava flow that threatened to destroy the fishing port was halted by spraying it with sea water.

Hazard mapping is valuable for predicting the flow paths of lahars and mudflows. Sediments deposited by previous lahars can be mapped to show, historically, the areas at greatest risk.

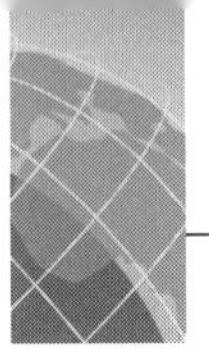

When an eruption is imminent, hazard warnings are broadcast to local populations and evacuation is ordered. Lahar detection systems around Mount Rainier in Washington state trigger automatic alerts and managers implement emergency evacuation for communities up to 100 km from the volcano.

What are the hazards associated with mass movement and slope failure?

Key ideas	Content detail
• Mass movement is more likely to occur when both physical and human factors disturb the equilibrium of a slope. • Mass movement has a number of impacts on the areas affected, which result in a range of human responses to the hazard.	• The processes and conditions that lead to mass movements: physical conditions (slope angle, weathering, vegetation, climate and weather, drainage, rock types) and human activities (including deforestation, adding weight, undercutting of slopes, quarrying). • The processes involved in the main types of mass movement: slides, flows, heaves. • Specific mass movement events: their causes; environmental, social and economic impacts; and the human response, both short term (emergency aid) and long term (planning and management).

Key questions

What are mass movements and mass movement hazards?

Mass movements are the downhill transfer of slope materials as a coherent body. Three broad types of mass movement are identified, based on their speed of movement and water content.

1 **Slides** are masses of material that move across a clearly defined slide plane. As a result the velocity is uniform throughout the sliding mass.

2 In **flows**, velocity decreases with depth. Typically, flows have a higher water content than slides and are therefore more mobile.

3 **Heaves**, such as soil creep and frost creep, occur slowly and in most environments pose little threat to life and property.

Mass movements become hazardous when they have a damaging effect on economy and society. If damage to property and/or loss of life is particularly high, a hazard becomes a **disaster**. The most hazardous mass movements are those that occur rapidly and with little warning — **debris flows**, **mudflows**, **mudslides** and **landslides**.

What physical and human factors trigger mass movements?

Mass movements occur on slopes where the driving force is gravity. Two sets of forces operate on slopes: **downslope** forces and **upslope** forces. Gravity is the main downslope force and increases with slope angle. The mass of mineral material and water on a slope also exert a downslope force. Upslope forces resist mass movement. They include the shear strength of slope materials, the frictional resistance between slope materials, and the binding effects of vegetation. Mass movement events are usually related to an external trigger that may be either physical or human:

- steepening of slopes by erosion or human activity (e.g. road or railway cutting)
- undercutting the foot of a slope (e.g. by a river erosion), removing basal support
- increased loading on slopes due to heavy and prolonged rainfall, building, tipping of waste materials
- heavy rainfall that lubricates slope materials and reduces frictional resistance
- heavy rainfall that increases pore water pressure and reduces the coherence of slope materials
- deforestation that (a) reduces the binding effect of tree roots on the slope materials, and (b) increases the amount of water absorbed by slope materials
- earthquakes or the vibrations caused by heavy lorry traffic or explosions (e.g. quarry blasting)

To what extent do human factors contribute to the causes of mass movement hazards?

Deforestation either by overgrazing or by deliberate logging for timber/fuelwood or new farmland, disrupts the balance of forces on slopes. Torrential rainfall can then trigger mass movements such as landslides, mudslides and mudflows. The mass movement disasters in Honduras (1998), northern Venezuela (1999) and Guinsaugon (Philippines, 2006), were all related to deforestation of steep slopes and extreme rainfall events. Mass movements, caused by torrential rainfall from Hurricane Mitch, killed thousands of people in Nicaragua in 1998.

However, some mass movements hazards are caused entirely by natural processes. In 1985, the eruption of Nevado del Ruiz in Columbia produced mudflows (or lahars) that killed 23,000 people in the town of Almero. Similar mudflows occurred around Mount Pinatubo in the Philippines, which erupted in 1991, and deposited huge ashfalls on the surrounding area. Even today, 20 years after the eruption, heavy rains create powerful mudflows that are hazardous to local people.

What is the relative importance of physical and human factors in causing mass movement hazards?

Unlike in earthquake and volcanic hazards, human activities can be a causal factor in mass movement hazards. They contribute to mass movements by destabilising slopes. The mass movement disasters of Guinsaugon and northern Venezuela were partly the result of deforestation, which triggered mudslides and debris flows.

Taiwan is highly susceptible to severe mass movement hazards. This susceptibility is due to a combination of physical and human factors. Physical factors include high rates of tectonic uplift, steep slopes, weak rocks, earthquakes and extreme rainfall events. In August 2009, Typhoon Morakot dumped 2,000 mm of rainfall in just 3 days causing landslides and debris flows that killed hundreds of people. But susceptibility to mass movements is also high because of development in the mountains. Road construction, the building of hotels and hot spring resorts, HEP schemes and fruit farming have degraded the environment and increased the risks from mass movement hazards.

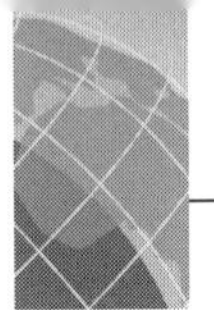

Some mass movement hazards, however, show no obvious causal connection to human activities. The Sichuan earthquake in central China (2008) and the Sumatran quake (2009) were extreme events that triggered hundreds of landslides, which destroyed property and killed thousands of people. In both disasters, mass movement hazards were not directly related to human activity.

How important are physical and human factors in determining the impact of mass movement hazards?

The impact of natural hazards in general can be explained by two factors: exposure and vulnerability. **Exposure** refers to the magnitude and frequency of the natural event and the number of people living in the areas of risk. Thus, in mountainous regions like Kashmir and Taiwan, with steep slopes, large populations and frequent earthquakes, exposure to mass movement hazards is high. **Vulnerability** is the preparedness of a population in relation to a particular hazard and its ability to respond and mitigate its effect. Vulnerability to mass movement hazards is lessened by programmes of reafforestation/forest conservation on steep slopes, hazard mapping of areas of highest risk, emergency disaster planning etc.

Analysis of the impact of most mass movement hazards reveals the importance of exposure and vulnerability. The northern Venezuela debris flows of 1999 killed an estimated 30,000 people because the area suffered both high exposure and high vulnerability. Steep slopes, high mountains, extreme rainfall intensified by relief, and large populations located on a narrow coastal plain and in the path of debris flows created high exposure. But the 1999 disaster was amplified by the vulnerability of the population. There was a lack of preparedness, with no strategy for managing a major natural disaster. No attempt was made to remove impromptu slums built on steep slopes near the coast, and no action was taken to clear sediments and debris, which clogged river channels in the mountains. Finally, widespread deforestation in the mountains accelerated run-off, increasing the risk of deadly mass movements.

How can the impact of mass movement hazards be mitigated?

Mitigating the impact of mass movements means making people and society less vulnerable to these natural events. There are two approaches: (a) to reduce the risks of mass movement occurring in the first place and (b) to respond to actual mass movement events to limit loss of life, injury and damage to property.

In mountainous environments, reafforestation can greatly reduce the risks of landslides, mudflows and other mass movements. Conservation management of this type is currently being implemented in the Annapurna region of Nepal. Local farmers are given an incentive to conserve forests and woodlands by receiving a proportion of the income generated by ecotourism. Steep hillsides that are cultivated and where mass movement is a threat, are often terraced. Terracing is common throughout the rice-growing areas of Asia, especially in Indonesia and the Philippines.

Artificial slope drainage reduces the risks of mass movements by lowering pore water pressure, increasing frictional resistance and reducing loading. Slopes that have been oversteepened in road and railway cuttings can be bolted, or shored up with metal pilings or basal walls. Other management responses to minimise the mass movement hazard include the publication of hazard maps, showing areas of risk (e.g. the paths followed by mudflows and debris flows). Early warning systems can also be put in place, giving time for orderly evacuations.

Once a mass movement disaster is a reality, disaster planning can get vital emergency aid to survivors, airlift the injured to hospital and organise search and rescue teams. Long-term development plans are also needed to reconstruct housing, infrastructure and local economies.

What are the hazards associated with flooding and why do the impacts on human activity vary over time and location?

Key ideas	Content detail
• Flood risk reflects a combination of physical and human factors, which vary from place to place. • Flooding has a number of environmental and social impacts on the areas affected, which result in a range of human responses to the hazard.	• Rivers and coastal floods are influenced by a range of physical factors (including height, relief, drainage regime, climate, vegetation and rock type) and by human factors (including settlement building, farming, deforestation and drainage). • The impacts of floods are environmental, social and economic. • The human reaction to floods is both short term (emergency rescue) and long term (planning and management).

Key questions

What physical factors influence river flooding?

Rivers flood when they overtop their banks and inundate the adjacent valley floor (**floodplain**).

The most obvious cause of flooding is intense and/or prolonged rainfall. Heavy **convectional rainfall** associated with thunderstorm cells causes **flash floods**, which are most common in the UK in the summer months. **Slow floods** occur after days or weeks of prolonged rain or snowmelt. They are less hazardous than flash floods as, being more predictable, they allow early warning.

Many rivers, particularly in the tropics and sub-tropics, have an uneven flow pattern during the year (**regimes**). In tropical Africa and monsoon Asia, rivers flood annually during the wet season. For the rest of the year, discharge is low and may cease altogether. Rivers with headwaters in glaciated mountainous regions are most likely to flood following the spring and early summer thaw.

Relief and geology also influence flooding. Not only do upland areas intensify rainfall events (**orographic effect**) but their steep slopes also cause rapid **run-off**, increasing **peak flows** and the flood threat (e.g. Cumbria floods, 2009). Where catchment geology consists of impermeable rocks (e.g. shale, clay), natural storage of water (i.e. **groundwater**) following rainfall events is minimal. As a result, a large proportion of rainfall becomes run-off and this again increases the flood risk.

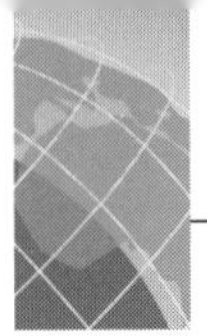

Vegetation has a direct effect on the amount of water reaching stream and river channels by (a) **intercepting** rainfall, part of which is then evaporated, and (b) absorbing water from the soil through root systems and **transpiring** it to the atmosphere. Where catchments are sparsely vegetated (e.g. arid environments), levels of interception and transpiration are low and rivers have a greater propensity to flood.

What human factors influence river flooding?

Human factors are often a causal agent in river floods. Flood hazards are increased by **urbanisation**, farming, **deforestation** and land drainage. Urbanisation modifies the local **hydrological cycle**. The fabric of towns and cities is dominated by impermeable materials such as concrete, tarmac, brick and tiles. These materials have little capacity to store water. Moreover, urban areas have efficient drainage systems (e.g. pitched roofs, gutters, drains and sewers) designed to remove water as quickly as possible. Together, impermeable surfaces and artificial drainage promote rapid run-off. Thus streams and rivers draining urban areas often have short lag times and high peak flows, which increase the probability of flooding. Furthermore, urban growth often extends onto floodplains, which are natural storage areas for floodwaters. The draining of floodplains to allow urban development can greatly increase the flood risk, especially in areas downstream.

Some farming practices also increase the risk of flooding. In mid-latitudes, replacing pasture with arable crops leaves fields with little plant cover for half the year, reducing interception and transpiration, and increasing run-off. This effect may be amplified by underdrainage. In the UK, many moorland environments were drained in the late twentieth century to improve the quality of pasture. One side effect was that rivers became more 'flashy' with shorter lag times and higher peak flows. Unsustainable farming practices (e.g. overcultivation, cultivation of steep slopes) may lead to **soil erosion** and the **siltation** of rivers. The build-up of silt in rivers reduces channel capacity and again increases the flood risk.

Deforestation significantly alters the movement of water in drainage basins, reducing interception and transpiration but increasing run-off. On steep hillslopes, deforestation is likely to result in soil erosion. Rivers that drain catchments that have suffered widespread deforestation (e.g. the Brahmaputra in South Asia) invariably show an increased frequency of flooding.

What physical factors influence coastal flooding?

Flooding occurs along lowland coasts due to exceptionally **high tides** or **storm surges** or a combination of both. Exceptional high tides, known as spring tides, occur twice a month and the highest tides of the year happen around the equinox. Coastal flooding is most likely at these times. Storm surges, whipped up by gale force winds and low pressure can raise sea levels several metres. The floods in New Orleans in 2005 in the wake of Hurricane Katrina were caused by an 8 m storm surge.

Climate also has an effect on coastal flooding. Coastlines in the sub-tropics that experience violent **hurricanes** and **tropical storms** face high risks from flooding. Similarly in middle and high latitudes, depressions, like hurricanes, are powerful storms that generate strong winds, high waves and elevated sea surfaces.

Coastal areas most susceptible to flooding are deltas, estuaries and reclaimed fens and marshlands. In the UK, lowland coasts such as the Broads in East Anglia and the Somerset Levels are at greatest risk from flooding.

Vegetation can lessen the threat of coastal floods. In the tropics, **mangroves** encourage sedimentation and reduce wave power. **Salt marshes** have the same effect in higher latitudes. A country like Bangladesh faces increasing risks of coastal flooding partly because of the destruction of its mangrove forests. Reclamation of salt marshes in the Mississippi Delta in the past century contributed to the devastating effects of the Hurricane Katrina storm surge.

What human factors influence coastal flooding?

Human factors are less significant as causal agents in coastal floods than in river floods. At the global scale, many lowland coasts are experiencing a heightened flood risk. This is due to rising sea levels linked to global warming. Today, most scientists agree that **global warming** is **anthropogenic** — the result of increasing levels of greenhouse gases in the atmosphere due mainly to the burning of fossil fuels and deforestation. Large parts of Bangladesh are already under serious threat from rising sea levels and may have to be abandoned within the next two or three decades. Global warming is also responsible for climate change and the increased frequency and power of storms such as hurricanes and mid-latitude depressions. Settlements in lowland coastal areas such as the Gulf of Mexico and the Bay of Bengal face an uncertain future as powerful storms like Hurricane Katrina and Cyclone Nargis (Myanmar, 2008) become more common.

In the Mississippi Delta natural gas extraction has caused widespread land subsidence, which has increased the flood risk to New Orleans. The drainage of salt marshes has had a similar effect, while the clearance of mangrove forests in Bangladesh has exposed coastal communities to greater flood risks.

Flood hazards arise because millions of people live in coastal regions that are prone to flooding. Rapid population growth in coastal communities on the Gulf of Mexico has increased levels of exposure to storm surges associated with hurricanes. Similar population increases have occurred in deltaic regions in Bangladesh and Myanmar.

What are the environmental, social and economic impacts of floods?

River and coastal floods have widespread environmental, social and economic impacts. In July 2009 floodwaters on the floodplain of the River Wear near Durham

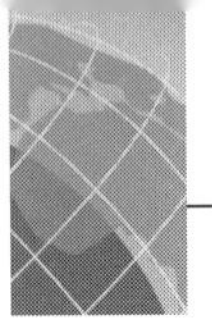

carved a huge gulley in a field of barley, removing 12,000 m^3 of soil and gravel in just a few hours. The July 2007 floods in the Severn Valley in Gloucestershire also caused significant environmental damage. Thousands of small mammals drowned, ground-nesting birds were badly affected and large numbers of fish were left stranded on the floodplain. Environmental damage also has adverse economic and social effects. The storm surge generated by Cyclone Sidr, which hit Bangladesh in 2007, badly damaged large areas of coastal mangrove (a source of timber and food), which could take 40 years to recover. Coastal floods in Bangladesh in 2009 led to the salinisation of farmland, which will remain uncultivable for years.

Major river and coastal floods cause death and injury. The 1931 river floods in China killed an estimated 2–4 million people. The combined death toll from floods on the Yellow River in China in 1887 and 1937 was well in excess of 1 million. Coastal flooding in the wake of Cyclone Nargis killed 140,000 people in Myanmar, while the storm surge that followed Hurricane Katrina caused 1,400 deaths. Severe coastal floods occurred in the Netherlands and eastern England in January 1953 — 1,800 people lost their lives in the Netherlands, while in eastern England 309 people died and 30,000 were evacuated to escape the floodwaters.

The principal economic effects of floods are damage to property and infrastructure. This tends to be higher in developed than in developing countries because of (a) greater fixed investments and (b) assets such as land, housing and commercial buildings with higher value. In addition, most assets are protected by insurance that makes good the losses following a flood disaster. The economic cost of Hurricane Katrina — over US$90 billion — was the costliest natural disaster in US history. Coastal flooding, especially in New Orleans, accounted for most of the cost. The total insured loss from the July 2007 floods in Gloucestershire was £1–1.5 billion. These losses covered damage to property and motor vehicles, as well as the disruption to businesses and the expense of providing temporary accommodation for people forced to leave their homes. There was also damage to crops submerged and contaminated by floodwater.

What are the short-term responses to flood disasters?

These focus on evacuation and emergency aid. Early warnings of floods allow time for evacuation to safe areas. In the Brahmaputra-Ganges delta in Bangladesh the death toll from storm surges since 1991 has been greatly reduced thanks to the construction of elevated cyclone shelters, and the implementation of effective early warning.

Major flood disasters such as Cyclone Nargis require immediate short-term aid. Flood victims need food, fresh water, temporary shelter, sanitation and medicines. Often the only way to reach survivors is by helicopter. Most developing countries lack the resources to respond effectively to major flood disasters and rely on the international community — foreign governments, NGOs and multilateral agencies — for emergency relief and aid.

What are the long-term responses to flood hazards?

Long-term responses to flood hazards fall into two categories: **structural** and **non-structural**. Structural responses rely on hard engineering to confine river floodwaters to river channels and protect coastal areas from high tides and storm surges. **Dams** and **reservoirs** store floodwaters which are released gradually and eliminate extreme flow events. **Sluice gates** and flood basins operate in a similar way. Gates are raised during a flood event and floodwaters are stored temporarily on the surrounding floodplain. **Flood embankments** or **levées** raise the height of river banks and increase channel capacity. The level of flood protection they provide depends on their height. **Channel straightening** works by increasing flow velocities, thus increasing the **bankfull discharge** of the river. Artificial channels, known as **flood relief channels**, divert a part of a river's flow and thus reduce discharge in the main channel. Flood relief channels are often built to protect particular settlements but provide no protection for places further downstream.

Structural responses which prevent coastal flooding include **seawalls**, **flood embankments** and **flood barriers**. Seawalls are designed to prevent overtopping and flooding by spring tides and storm surges. Expensive to construct, seawalls are only built where large settlements and important infrastructure are at risk. On rural stretches of coast, flood embankments replace seawalls. In some estuaries, where the tidal range is amplified, flood barriers (e.g. Thames barrier) protect major centres of population. The flood gates within the barriers are closed during exceptionally high tides and then opened on the ebb tide to release water from the landward side.

Non-structural approaches to flooding cover **flood abatement**, planning controls and flood management. Flood abatement aims to prevent floods developing in the first place. Afforestation of catchment headwaters increases lag times, reduces peak flows and is effective in limiting the size river floods. Comparable measures in coastal regions include the conservation of mangroves in the tropics, and salt marshes and mudflats in middle and high latitudes. Controlling development in areas of highest flood risk (e.g. floodplains) is increasingly favoured. Climate change and forecasts of more frequent floods makes this approach more sustainable. It is supported by the production of detailed hazard maps (based on satellite imagery) that show areas most at risk from floods.

Flood warnings issued by national agencies help to mitigate the impact of floods. In Bangladesh, cyclones and tropical storms are monitored by the National Warning Forecasting Centre in Dhaka. The agency issues warnings 24 hours in advance. The most serious warning — 'great danger' — is issued 10 hours in advance.

Finally, in developed countries the impact of flooding on businesses and individual households is mitigated by flood insurance. Insurance can also help to reduce losses to flood damage by offering lower premiums where, for example, ground floors are reserved for garage spaces and where building materials are used that are less easily damaged by floodwater.

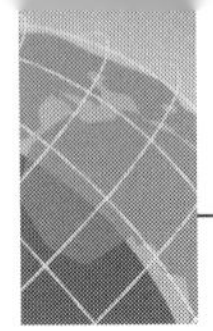

Ecosystems and environments under threat

What are the main components of ecosystems and environments and how do they change over time?

Key ideas	Content detail
• Ecosystems and environments are systems in which a number of components (physical and human) interact. • Ecosystems and environments are subject to constant change as the physical conditions and human activities operating upon them change.	• The concept of open and closed systems. • The interconnections between stores and flows in ecosystems. Stores and flows include energy and nutrients. • How change occurs in ecosystems and environments as a result of the interaction of physical and human factors.

Key questions

What are ecosystems?

Ecosystems are communities of plants, animals and other organisms and the physical environment that supports them. The physical environment comprises rocks, soils, water, solar energy, the atmosphere, fire and gravity. Interaction between the living (**biotic**) and non-living (**abiotic**) parts of ecosystems binds the two components together. As a result they function as coherent and **interdependent** wholes. This quality of interrelatedness, which causes ecosystems to behave like a single organism, is known as **holisticity**.

In what sense are ecosystems 'open'?

Ecosystems are open systems because their boundaries allow the passage of both energy and materials. The main energy input is short-wave solar radiation which is balanced by losses of long-wave terrestrial radiation or heat. Ecosystem boundaries are also open to inputs and outputs of materials including water, minerals and living organisms.

How is energy transferred within ecosystems?

Radiant energy from the sun is 'captured' by green plants and converted to chemical energy through **photosynthesis**. Green plants or **autotrophs** are therefore the primary energy producers in ecosystems. We measure the amounts of energy produced by autotrophs in grams per square metre per day or per year. This **gross primary production** depends on sunlight, temperature, water supply and nutrient availability. A more accurate measure of productivity is **net primary production**, which takes account of the energy consumed by **respiration**.

Green plants represent the first stage (or first **trophic** level) in the food chain. All other organisms in ecosystems (e.g. animals and decomposers) are **consumers** and depend ultimately on this primary production of energy. Animals that feed directly on plants are **herbivores** and occupy the second trophic level. Most herbivores are insects, though the best known are hoofed grazers (ungulates) such as antelopes, buffalo and horses. Meat-eating animals or **carnivores** occupy the third and higher trophic level. Carnivores include a wide range of species from insects (e.g. wasps) to fish (e.g. sharks) and mammals (e.g. wolves). Many animals are opportunists and feed on both plants and other animals. Well-known **omnivores** include creatures such as grizzly bears, warthogs and badgers.

Omnivorous animals remind us that the concept of food chains is a gross simplification. In reality a nexus of complex energy flows known as a **food web**, rather than a simple linear flow of energy, provides a more accurate model of energy transfers in ecosystems.

Dead plants and animals are ultimately consumed and decomposed by organisms (e.g. fungi, bacteria) or **detritivores**. Detritivores perform the vital function of mineralising organic material or nutrients and returning them to the soil.

There is a reduction in the amount of energy at each successive trophic level in a food chain. This is due to energy loss through respiration. Most energy consumed by an animal is not converted to body tissue, but is used up in vital metabolic processes and dissipated as heat. For this reason, food chains are limited in length to four of five trophic levels. The **biomass** at each trophic level (dry weight of organisms per unit area) falls rapidly with distance from the site of primary production. The result is a **pyramid of biomass**, dominated by plants.

How are nutrients cycled within ecosystems?

Nutrients are the chemical elements and compounds required by living organisms. Green plants need 16 essential nutrients. They include **macro-nutrients** needed in large quantities, such as nitrogen and phosphorus, and **micro-nutrients** like copper, zinc and iron. The ultimate source of most nutrients is rock weathering, which releases nutrients to the soil. However, some nutrients are also sourced in gaseous form from the atmosphere (e.g. carbon) and some are present in rainwater.

Unlike energy, ecosystems rely on a finite supply of nutrients, which have to be recycled. Nutrient stores include the soil, living plants and animals, dead organic matter (**litter**) and the atmosphere. The size of nutrient stores varies between ecosystems. In the tropical rainforest, where nutrient recycling is rapid, forest trees are the main nutrient store. In the boreal coniferous forest, where low temperatures slow down decomposition, the litter layer on the forest floor is an important nutrient store.

Nutrient cycling begins when plants remove nutrients from the soil through their roots. These nutrients, together with carbon absorbed in photosynthesis are then stored in the **phytomass** (plant biomass). Some nutrients are transferred to animals that graze or browse the vegetation. Eventually, nutrients return to the environment

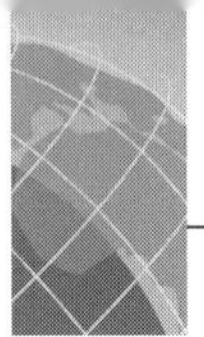

as dead organic matter which is decomposed by fungi and bacteria. The nutrients are then recycled.

Under natural conditions, nutrient cycles are virtually closed, with very little leakage. Any nutrients lost through rainfall by **leaching** are balanced by fresh nutrient inputs from rock weathering.

How does natural change occur in ecosystems?

Ecological succession describes the process of natural change in ecosystems. **Primary plant succession** occurs in environments not previously vegetated, such as sand dunes and lava flows. **Secondary succession** is found on sites where the original vegetation cover has been destroyed (e.g. by wildfire).

Ecological succession follows a number of stages or **seres**. Each sere modifies the environment (e.g. by increasing soil depth and moisture availability, reducing exposure etc.), allowing new plant species to invade and achieve dominance. The initial seres mainly comprise **pioneer species** that can survive in extreme environments with little soil or moisture. Classic pioneer species include marram grass in coastal dune environments and lichens and mosses on bare rock surfaces. Over time the physical environment ameliorates, and **biodiversity**, biomass and primary productivity increase. The final stage is when a plant community achieves equilibrium with the environment and undergoes no further change. This stage of succession, known as **climax**, is usually controlled by climate (e.g. tundra, boreal forest). Sometimes local environmental factors such as soils, geology and slopes influence the climax form: this is called a **sub-climax**. Some climax forms are in balance with both the physical environment and human activities. Plant communities of this type are known as **plagio-climaxes** (e.g. lowland heaths and heather moorland).

How does the interaction between people and the physical environment affect ecosystems?

Human activities modify natural ecosystems. These activities may be deliberate (e.g. firing grassland to improve the quality of pasture) or inadvertent (e.g. introducing alien species) but their impact is invariably negative. The main ecological effects are:

- a reduction in biodiversity
- a reduction in energy flows
- a shortening of food chains and a reduction in the complexity of food webs
- a reduction in net primary production and biomass
- greater instability due to the reduction in ecosystem complexity and structure, and therefore in the number of negative feedback loops

In many hot arid and semi-arid environments, the human impact on ecosystems has been disastrous. In the Sahel in Africa and throughout much of the Middle East, many healthy ecosystems have been degraded by overgrazing, overcultivation and deforestation, creating desert-like conditions.

What factors give particular ecosystems and environments their unique characteristics and in what ways are they threatened by human activity?

Key ideas	Content detail
The interaction between physical and human factors creates distinctive ecosystems and environments and leads to change within them.	The study of a local ecosystem (coastal dunes) to include: • the main stores and flows • the physical factors that influence dune ecosystems (microclimate, soil, relief, drainage) and how the ecosystem develops with time • the main human influences and how they generate change with time • the threats posed by human activity

Key question

How does the interaction of physical and human factors lead to change in coastal dune environments?

Coastal dunes develop along lowland coastlines, where prevailing winds blow onshore, offshore gradients are shallow and there is abundant sand. Dunes are fragile environments. Unaffected by human activities, a delicate balance exists between vegetation, wind speed and sand movement. Pioneer species such as marram and sand twitch, adapted to dry and exposed dune environments, trap wind-blown sand and drive the growth of dunes. They also modify dune habitats by providing organic material and nutrients for soil development, moisture retention and shelter. As succession occurs, biodiversity, biomass and primary production increase. Meanwhile the dunes become anchored and develop as a series of ridges parallel to the shoreline. The oldest dunes are located furthest from the shoreline.

Today, many coastal dune systems are under pressure from human activities such as recreation and leisure, housing developments, forestry, coastal defence schemes and alien plant species. At Sefton on Merseyside, uncontrolled access to the dunes has damaged the vegetation through trampling and firing and has led to severe erosion. Activities such as off-road driving have accelerated this process. Fifty per cent of the dunes have been lost to housing and building, and a road that cuts through the dunes has halted the natural dune-forming processes. Links golf courses now occupy one quarter of the area of the dunes. Dune habitats have also been lost to afforestation, which modify soils (more acidic) and microclimates. Seawalls to the north and south of the Sefton dunes protect the Southport and Crosby areas from erosion, but reduce the natural sand supply required to build the dunes and keep them healthy. Finally, introduced species such as sea buckthorn and white poplar, planted to help stabilise the dunes, have colonised large areas, causing nutrient enrichment.

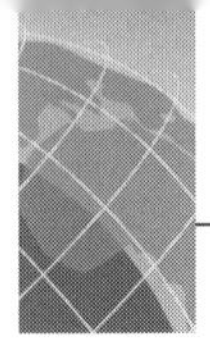

Why does the impact of human activity on the physical environment vary over time and with location?

Key ideas	Content detail
The impact of human activity on environments varies between different areas of the world at different stages of economic and technological development.	Countries at different ends of the development continuum have: • variable impacts on the physical environment (both positive and negative) • impacts on the physical environment which can either increase or decrease with economic, social and technological development

Key questions

How do countries at opposite ends of the development continuum impact the physical environment?

Kenya, a poor country in east Africa, is at the lower end of the development continuum. Its main economic activities are farming and tourism. In the highlands in central Kenya, large-scale agribusiness has transformed the natural environment into plantations and large estates for growing export crops such as coffee, tea, flowers and fresh vegetables. Elsewhere, farming is small scale and semi-subsistence. Livestock farming predominates in the drier areas in the north and west, where there is sufficient rainfall for cultivation. Rapid population growth and overstocking has led to **land degradation** and **desertification** in many parts of Kenya. Wild animal populations have declined due to competition for grazing and water with domestic cattle and goats. Overgrazing has produced severe soil erosion, especially in the more arid areas.

Tourism has had a more positive impact on the environment. Kenya's major tourism attraction is its spectacular wildlife, particularly its large mammalian herbivores (wildebeest, buffalo, antelope) and predators (lions, leopards, hyenas). Kenya's wildlife is protected by a system of well-established National Parks (e.g. Ambroseli) and game reserves (e.g. Masai Mara). As far as possible, these areas maintain the natural **savanna** ecosystem. But even in the parks and game reserves there are threats to the environment. They include poaching of larger animals such as elephant and rhino; habitat loss due to illegal grazing of domestic livestock by indigenous pastoralists; and disturbance to wildlife and erosion caused by tourists and their vehicles on safari.

Rich countries like the UK also have both positive and negative effects on the environment. Conservation of landscapes, habitats and wildlife is the priority of the National Parks, which cover approximately 10% of England and Wales. Other conservation areas include Areas of Outstanding Natural Beauty (AONB), Nature Reserves and Sites of Special Scientific Interest. Although the degrading effects of human activities in National Parks and AONBs are relatively small, conservation has to be balanced with maintaining/providing jobs for local communities. Outside conservation areas, environmental protection has, until recently, had little priority. Thus in the past 60 years there have been dramatic reductions in areas of ancient woodland, traditional hay meadows, chalk downland, hedgerows and wetlands. Intensive farming and the excessive use of **agro-chemicals** are largely responsible for loss of habitat, the pollution of streams and rivers and the steep decline of farmland bird populations.

Today conservation has a higher priority. Thanks to a wide range of policies from the EU and the British government, farming has become more environmentally friendly and sustainable.

How do impacts of human activity on the physical environment change with economic, social and technological development?

The environmental impact of human activity increases in scale with economic, social and technological development. Indigenous cultures, based on simple technologies have stable populations and sustainable economies, causing no lasting damage to ecosystems. Examples include **shifting cultivators** in the rainforests of South America and central Africa, reindeer herders like the Sami in northern Scandinavia, and **hunter-gatherers** such as the San of the Kalahari. With simple technologies, these groups have a limited ability to change their environment. Moreover, because they rely on the resources of local ecosystems, the need for sustainable living is essential to their survival.

Increasing technology brings greater control over the environment. It also stimulates population growth, which puts pressure on the environment. In Europe the combination of advanced technology, wealth and high population density has led to drastic modification or outright destruction of many natural ecosystems. For example, few areas of primary deciduous woodland survive, lowland areas have been deforested and replaced by intensive farming, wetlands have been drained, and upland valleys flooded by reservoirs. The massive loss of habitat accompanying these changes has fragmented plant and animal populations, destroyed food webs and reduced biodiversity. Even in poor countries, traditional societies have been affected by imported technologies that have had adverse implications for the environment. Improved mortality control has triggered rapid population growth, which has often been the driver behind the overexploitation of natural resources and land degradation.

The most profound environmental change associated with economic development is caused by urbanisation. In urban areas, the natural environment undergoes profound transformation. Vegetation is cleared and soils are covered by tarmac and concrete; **microclimates** and **energy exchanges** are modified by artificial surfaces, heat production, pollution and buildings; and the hydrological cycle is fundamentally altered, with changes to evaporation, transpiration and run-off.

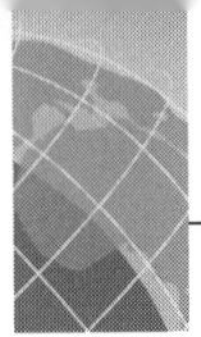

How can physical environments be managed to ensure sustainability?

Key ideas	Content detail
Human activities which impact adversely on physical environments need to be managed in order to be sustainable.	Sustainable environmental management which includes conservation, planning, controls and restricted use.

Most National Parks in England and Wales have been established for over 50 years. Their primary purpose has been to protect and conserve areas of countryside of outstanding natural beauty, while at the same time promoting public access for recreation and leisure, and allowing local economies to flourish. Where conflict arises, primacy is given to conservation.

Conservation in these National Parks has been a resounding success. Thanks to conservation, classic landscapes such as Derwent Water in the Lake District and Upper Wharfedale in the Yorkshire Dales are little altered from the way they looked 50 years ago. This is consistent with the idea of **stewardship** and has been achieved despite huge increases in visitation to National Parks. This sustainable use of National Parks requires careful management and planning. The parks receive an annual grant from central government and are managed by National Park Authorities (NPAs). Funding is 'topped up' from parking fees, planning applications and sales to tourists at visitor centres.

The NPA is the sole planning authority in the National Parks. It is guided by principles of conservation and enhancement of the environment, sustainability and the economic viability of local communities. Its main conservation tool is planning control to prevent inappropriate development. It includes decisions on the location of new infrastructure, new buildings, and building styles and materials. In 2005, the Lake District NPA imposed a 5 mph speed limit on boats and other craft on Windermere, which stopped noisy water-based recreation such as water-skiing and jet-skiing. NPAs also maintain footpaths and public rights of way, and zone land use to avoid conflict. The most isolated areas in the Lake District National Park are reserved as quiet, eco-zones, while intensive recreation is concentrated in the more accessible areas around Keswick, Ambleside and Windermere, where planning control is more relaxed.

Other public agencies influence conservation in National Parks. Farmers who adopt sustainable and eco-friendly methods and who maintain landscape features such as stone walls, hedgerows and field barns, receive grants from DEFRA. The Common Agricultural Policy provides subsidies for upland farmers who agree to reduce stocking levels in line with the land's carrying capacity. Meanwhile, bodies like the Environment Agency tackle pollution problems on rivers and lakes. In the Lake District, the National Trust is the largest single landowner. It looks after places of historic interest and natural beauty for the benefit of the nation.

Climatic hazards

What conditions lead to the development of tropical storms and tornadoes, and in what ways are they hazardous to people?

Key ideas	Content detail
• Tropical storms and tornadoes form and develop under particular atmospheric conditions. • Tropical storms and tornadoes are hazardous and have serious environmental, social and economic impacts upon the areas they affect.	• The atmospheric and surface conditions that give rise to tropical storms and tornadoes, and the processes involved in their growth and development. • The types of hazards presented by tropical storms (winds, storm surges, flooding etc.) and tornadoes (violent winds) and the impacts these hazards have.

Key questions

How do tropical storms form and develop?

Viewed from above, tropical storms (also known as hurricanes, tropical cyclones and typhoons) comprise numerous curved bands of **cumulo-nimbus** clouds that take on a familiar circular form. At the centre of the storm there is 10–65 km cloud-free area of light winds called the **eye**. Tropical storms develop over tropical oceans between latitudes 8° and 20°. Their formation depends on three conditions:

1 Plenty of water vapour.
2 Light winds to allow vertical cloud development.
3 **Sea surface temperatures** of at least 26–27°C.

These conditions are found in late summer and early autumn in the tropical North Atlantic Ocean, the Bay of Bengal and Arabian Sea, and the North Pacific Ocean.

Tropical cyclones begin life as clusters of thunderstorms over the ocean. Some of these clusters become better organised and develop a **cyclonic** circulation in response to the Earth's rotation. Surface pressure falls as **condensation** releases **latent heat** which warms the atmosphere and causes uplift. The result is a feedback process: atmospheric warming induces more **evaporation**, more condensation and more latent heat. As the fall in central pressure deepens, surface air is sucked in, ensuring an abundant supply of water vapour. Once sustained wind speeds reach 119 km h^{-1} the storm is classified as a hurricane/tropical cyclone/typhoon.

Tropical storms may take weeks to develop, but often disappear in just a few days. Rapid decay occurs when the storm loses its energy supply (i.e. warm, moist air). This happens when it either makes landfall or moves over an area of ocean with lower surface temperatures.

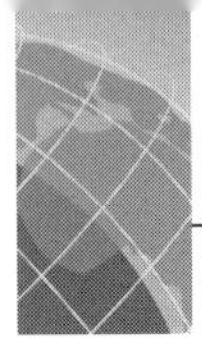

What are the hazardous impacts of tropical storms?

The primary hazards associated with tropical storms are hurricane-force winds, storm surges and torrential rain, the last causing river floods and mass movements.

The most powerful tropical storms are category 5, with sustained wind speeds in excess of 250 km h^{-1} and gusts up to 360 km h^{-1}. Storm damage results from both direct impact and flying debris. Tall buildings are vulnerable to collapse and entire forests can be flattened by hurricane-force winds. Most destruction, death and injury is due to flying debris, such as poorly fastened roof tiles, telegraph poles and other projectiles.

Storm surges are a bigger hazard and in lowland areas close to sea level can cause thousands of deaths. In the USA in 2005, Hurricane Katrina killed over 1,400 people, while Cyclone Nargis caused one hundred times more deaths in Myanmar in 2008. Storm surges are formed by high tides combined with powerful winds and low pressure, which raises sea levels that can overwhelm low-lying coasts and cause massive destruction. The economic impact of storm surges is huge. Hurricane Katrina is estimated to have cost over US$90 billion.

Slow-moving tropical storms dump enormous amounts of rain and trigger widespread river flooding. In October 1998, Hurricane Mitch hit Central America and brought severe floods and landslides. Between 1,000 and 2,000 mm of rain fell in the mountains of Honduras. Thousands of people were swept away by the floodwaters; debris flows, mudslides and other mass movements destroyed whole villages; and there was widespread destruction of infrastructure and food crops.

How do tornadoes form and develop?

A tornado is a violently rotating column of air that extends from the cloud base of a powerful thunderstorm (**supercell**) to the ground. This vortex forms a funnel visible as dust and other debris sucked up from the ground and water droplets. Within the funnel, wind speeds may reach 500 km h^{-1}. Compared to tropical storms, tornadoes are more localised and, on a small scale, even more destructive.

Tornadoes develop when two **airmasses**, one cold and dry and the other warm and moist, meet. The warmer, moister air lies below the colder, drier air. Steep temperature and pressure gradients develop aloft, and the atmosphere becomes highly **unstable** and creates powerful **updraughts** of warm air (**mesocyclone**). Intense low pressure within the mesocyclone causes air to rush in and spiral upwards. Fierce winds and low pressures exert enormous forces on objects in the path of the tornado, causing structures such as buildings to implode.

What are the hazardous impacts of tornadoes?

Tornadoes occur in all continents except Antarctica, but are most common in continental interiors like the US Midwest. The main hazards are violent winds and flying debris. In the most powerful tornadoes (level 5 on the Fujita tornadic damage scale), winds are strong enough to pick up cars and tear houses to shreds. Damage paths can be up to 1.5 km wide and 80 km long. In April 2006, a supercell brought massive tornadic damage to communities in eastern Kansas, Arkansas, Missouri and northwest Tennessee. Tornadoes rated 3 on the Fujita scale flattened trees, knocked down power lines and demolished homes and grain silos. In total the tornadoes killed 25 people and injured 176. Much of the town of Marmaduke in Arkansas was destroyed.

How do atmospheric systems cause heavy snowfall, intense cold spells, heatwaves and drought, and in what ways are they hazardous?

Key ideas	Content detail
• Anticyclones and depressions can produce extreme weather conditions which are hazardous. • Hazards associated with anticyclones and depressions have serious environmental, social and economic impacts on the areas they affect.	• Anticyclones and depressions and the formation of potentially hazardous events, i.e. heavy snowfall, cold spells, heatwaves and drought. • How these events represent hazards to people in the form of blizzards, cold spells, heatwaves (e.g. wildfires) and droughts (e.g. water shortages). • The impacts associated with these weather features at local, regional and global scales, including impacts on transport, agriculture and forestry, health and economic activity.

Key questions

How do anticyclones and depressions give rise to extreme weather events?

Anticyclones are areas of high pressure. In mid-latitudes, anticyclones are mobile but slow-moving features which may persist for days or even weeks. When a large persistent 'high' becomes established over western Europe, it **blocks** the normal rain-bearing westerly winds. The result is minimal rainfall, which may lead to drought. Settled conditions — a characteristic of anticyclones — can produce unusual amounts of sunshine or cloud. In western Europe in summer, clear skies and intense **insolation** occasionally brings **heatwave** conditions. Heatwaves pose serious health risks to the old and very young, and increase the hazard of **wildfires**. Winter anticyclones often result in sub-zero temperatures, severe night frost and prolonged cold spells. Depending on wind direction and humidity, there may be continuous sunshine or overcast, gloomy conditions.

Depressions are mid-latitude storms that dominate the weather in northern and western Europe. They develop on the **jet stream**, generally follow a northeasterly track across the North Atlantic and bring unsettled wet and windy conditions at all seasons. Long spells of Atlantic storms are a feature of the climate of western Europe (e.g. November 2009). When such spells occur in summer, caused by the jet stream pushing much further south than normal, they bring above-average rainfall. An example of this type of extreme event was the record rainfall in the UK in June and July 2007. In Yorkshire, rainfall in June was three times the average and in

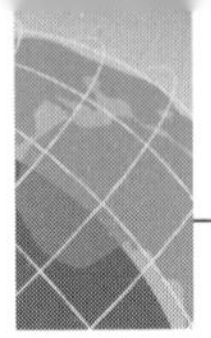

Worcestershire, July's rainfall was four times greater than average. The result was widespread flooding.

In winter, depressions with their frontal systems and strong winds can create blizzard conditions. In early March 2006, low pressure over Scandinavia and high pressure in the Atlantic Ocean produced a cold northerly airflow. Within this flow several small polar 'lows' developed, producing blizzards and heavy snowfalls. Snow storms spread across Scotland, forcing the closure of schools, major roads, railways and Inverness airport. At Aberdeen, 26 cm of snow was recorded on 3 March.

What are the impacts of droughts and heatwaves?

Droughts and heatwaves are hazardous and have environmental, social and economic impacts. Drought is an abnormal deficiency of rainfall over an extended period of time. In western Europe, drought has several adverse effects. As water tables fall and the soil dries out, river flows dwindle, lowering oxygen levels, increasing the concentration of pollutants and threatening aquatic life. Drought is a disaster for farmers who rely on direct rainfall. In the worst cases crops may die — more often crop yields are reduced. The drought that hit West Yorkshire between April and September 1995 caused reservoirs serving some towns to run dry. Domestic water supplies were only maintained by road tankers bringing water in from Northumberland. The consequences of drought are most severe among subsistence and semi-subsistence farmers in poor countries where crop failure and the death of livestock may result in famine, starvation and the complete break-up of communities. This happened in Niger in 2005, when, after 13 months of drought, 3 million people faced famine and severe food shortages.

Summer droughts are often accompanied by heatwaves, which pose significant hazards to human health. The heatwave that struck Europe in July and August 2003 caused 35,000 additional deaths, most of them elderly people. Mortality was highest in large urban areas where temperatures were up to 10°C higher than in the surrounding countryside. Heatwaves and drought are also responsible for wildfires such as those in Greece in August 2009, which destroyed 150 km^2 of forests around Athens.

What are the impacts of blizzards and cold spells?

Blizzards are periods of heavy snowfall accompanied by strong winds. Drifting snow is a particular hazard of blizzard conditions and leads to the closure of roads and railways. In late February and early March 2006, 4 days of blizzards brought much of northern Scotland to a standstill. In Shetland, drifts were up to 2 m deep. Hundreds of motorists were stranded and their vehicles abandoned on the A96 between Inverness and Aberdeen. Disruption to road traffic led to the suspension of domestic refuse collection in Aberdeen; several trains got stuck in drifts between Dundee and Aberdeen; flights from Aberdeen airport were severely disrupted; and most Scottish football fixtures were postponed. Meanwhile, schools in Shetland were closed on 4 successive days.

Prolonged cold spells in winter, especially when they are caused by anticyclones, do not always bring snow. While such dry cold spells will cause little disruption to the economy, low temperatures pose a serious health risk to old people with inadequate home heating. In Ireland, excess deaths during the winter months are 19%, due largely to poor housing and heating. In the UK poor people get financial assistance (£25 week) with heating bills when the average temperature is sub-zero on 7 consecutive days.

What are the impacts of heavy rainfall?

Heavy rainfall can lead indirectly to the pollution of rivers and coastal waters. In order to avoid flooding, storm water and raw sewage is diverted by combined sewer overflow pipes from sewage treatment works and is discharged into rivers and coastal waters. The resulting pollution is a health hazard to people and a threat to wildlife.

The main impact of heavy rainfall is flooding. **Flash floods**, caused by torrential convectional thunderstorms, are most common in summer. Recent examples are Boscastle (August 2004) and North York Moors (June 2005). Although the impact of flash floods is highly localised, they can be deadly. In August 1996 a flash flood which swept through a campsite at Biescas in the Pyrenees killed 71 people.

Slow floods, due to prolonged and exceptionally high rainfall, hit parts of England in June and July 2007, causing economic disruption and damage. Farming and tourism suffered the worst impacts. In East Yorkshire, 60% of the pea crop was destroyed and much of the potato crop rotted in the ground. Elsewhere, flooded grasslands reduced the silage crop for livestock feed and badly affected wheat yields. Shortages of potatoes, peas and cereals led to steep price increases in the supermarkets in the winter of 2007–08. Around Gloucester and Tewkesbury there was disruption to power and water supplies and total insured losses were £1–1.5 billion.

The wet summer also affected tourism. International tourist numbers to the UK fell in the summer of 2007. In Gloucestershire and Worcestershire large areas of countryside were closed because of flooding and thousands of British families booked last-minute holidays to southern Europe, affecting traditional UK seaside resorts like Brighton, Torquay and Scarborough.

Why do the impacts of climatic hazards vary over time and location?

Key ideas	Content detail
• The impact of climatic hazards on an area is influenced by levels of economic and technological development and population density. • The impacts of climatic hazards vary over time from the immediate to the long term.	• There are contrasts in the impacts of climatic hazards in countries at either end of the development continuum, rural and urban areas, coastal and inland areas. • The impacts of tropical storms and droughts vary over short and long time periods.

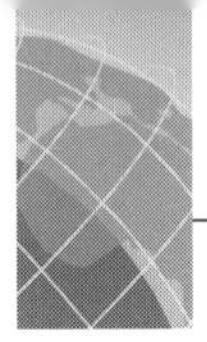

Key questions

How do climatic hazards impact countries at opposite ends of the development continuum?

Climatic hazards such as tropical storms and droughts have the greatest impact in poorer countries. Poorer countries are particularly vulnerable because they are less well prepared to cope with natural hazards. In addition, widespread poverty means that most individuals lack the resources and economic entitlements to buffer them against natural disasters.

Hurricane Katrina caused massive economic damage in the USA in 2005 (US$90 billion) and showed that even the richest countries are not immune to natural disasters. However, the death toll — 1,400 — was relatively low. This was due to effective hazard mitigation which reduced the vulnerability of the population of the Gulf states. They included advance warning of the hurricane's approach, evacuation strategies, disaster planning and the construction and maintenance of levées. This level of disaster mitigation is costly and confined to wealthy countries. Moreover, evacuation is easier in a country where most people own cars, telecommunications are universal and the road network is modern and efficient.

In May 2008, Cyclone Nargis hit Myanmar, one of the poorest countries in Asia. A storm surge, similar to Katrina's, killed 140,000 people and either destroyed or damaged 800,000 homes. The human cost of this disaster was, therefore, far greater than Hurricane Katrina. Most people hit by the disaster were poor farmers, who lost their crops and livestock and lacked the resources to survive without emergency relief aid. Cyclone Nargis highlighted Myanmar's vulnerability to storm surges, especially in the densely populated Irrawaddy Delta. Its impact was increased by poorly maintained levées along the coast, a lack of coastal radar to estimate the height of storm surges, and the failure of the government to provide warning until 24 hours before the storm made landfall.

Drought has a much greater human and economic impact in poorer countries. In 2005 the drought in southern Spain was the worst in 60 years. It caused severe water shortages, which damaged tourism and agriculture, lowered water tables and triggered wildfires, all of which had a major economic impact. The losses suffered by agriculture were €2–3 billion. Even so, in rich countries the impact of drought can easily be absorbed and for many might be little more than an inconvenience. But in poor countries, where most people rely heavily on subsistence and semi-subsistence farming, the human cost of drought can be devastating. Water holes dry up, animals die and crops fail. Indigenous people face food shortages and famine. In the summer of 2005 in southern Niger, drought reduced the grain harvest by a quarter, bringing famine to 3 million people. Despite international relief aid, thousands died from starvation and disease.

How do climatic hazards impact urban and rural areas differently?

The impact of extreme weather events such as heatwaves often varies between urban and rural areas. Urban microclimates are warmer than those in rural areas because:

- building materials such as brick, tarmac and concrete store heat in the day and release it at night
- air pollution helps to keep heat in
- there is less vegetation to cool the atmosphere
- cities produce heat (motor vehicles, space heating etc.)

In July and August 2003 an exceptional heatwave struck western Europe. It raised mortality rates sharply in large urban areas, where temperatures were several degrees higher than in the countryside. The heatwave caused 15,000 deaths in France, with around one-third in the Paris region alone. Rural areas, dependent on agriculture, are more likely to suffer economically in a prolonged heatwave (accompanied by drought) because of reductions in crop yields. Thus the 2003 heatwave badly affected wheat yields, which were 13% below average in Italy and 20% below average in France.

Rural areas are often more badly affected by climatic disasters such as tropical storms. This is partly because it is more difficult to access survivors and provide emergency aid. Bridges and roads, destroyed by floods and landslides, may leave many rural communities isolated in the days immediately after a storm disaster — and without food, clean water and medical supplies. Moreover, rural economies, dependent on farming, are more vulnerable to climatic hazards than urban economies, and urban populations (especially in developing countries) are on the whole better-off than rural populations, therefore more resilient in the face of climatic hazards and disasters.

To what extent do climatic hazards impact coastal and inland areas differently?

Coastal regions in the tropics and sub-tropics are exposed to tropical storm hazards. This is partly explained by the nature of tropical storms, which originate over the ocean and derive their energy from warm surface waters. Once tropical storms make landfall, they quickly lose their energy. Coastal regions are also, by definition, at or near sea level and so at particular risk from storm surges generated by tropical storms. Finally, coastal areas tend on average to be more densely populated than inland areas, further increasing their exposure to tropical storms and other climatic hazards.

In contrast, exceptional rainfall events that trigger mass movements are more likely to occur inland, and especially in mountainous regions where rainfall is intense and slopes are steep and often unstable. Much of the destruction associated with Hurricane Mitch in Central America in 1998 was the result of mudflows and mudslides in the mountains.

In the UK, blizzard conditions occur more often inland than on the coast. There are two reasons for this. First, mountainous regions like the Scottish Highlands increase precipitation through uplift and experience higher average wind speeds than lowland areas. Second, coastal areas experience milder conditions in winter, when sea surface temperatures are 2 or 3°C higher than those over the land.

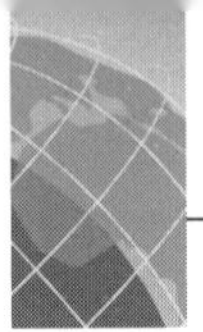

Although tornadoes can occur over the sea, they are more common and powerful in central continental areas like the US Midwest. Again, the explanation lies in the conditions needed for tornado formation, namely a humid warm airmass that develops over the ocean and tracks inland, and a dry, cool airmass of continental origin. The convergence of these two contrasting airmasses most often occurs in continental locations, hundreds of kilometres from the coast.

How do the impacts of climatic hazards vary over short and long time periods?

The immediate impacts of natural hazards such as tropical storms and droughts create emergency situations such as food shortages, lack of clean water, and the need for urgent medical treatment and temporary shelter. Resources to address these problems are provided by disaster relief agencies such as the Red Cross and the Federal Emergency Management Agency (FEMA), non-governmental organisations (NGOs) like Oxfam, and multilateral organisations (e.g. the UN). Governments focus initially on meeting essential needs and protecting life. Thus in the drought and subsequent famine in Niger in 2005, the priority was food aid, clean water and medicines. In the Hurricane Katrina disaster, the immediate need of the victims of flooding in New Orleans was accommodation. Thousands were housed temporarily in the city's Convention Center and in the Louisiana Sports Dome.

Natural disasters also have long-term effects: economies need to be reconstructed, infrastructure rebuilt and jobs provided. In New Orleans a decision has still to be made about the future of some low-lying suburbs several metres below sea level. Meanwhile, long-term reconstruction is under way. Levées high enough to withstand storm surges generated by category 5 hurricanes will be completed by 2011, and new pumping stations are being built. However, 4 years after the disaster, thousands of people had still not returned to their homes.

The long-term impact of severe drought in sub-Saharan Africa is often permanent land degradation and the destruction of pasture, water and soil resources, which once sustained whole communities. In extreme cases, farmers are forced to abandon the land and head for the nearest city. For those farming communities that remain viable, the legacy of drought might be long-term aid in the form of tube wells to guarantee water supplies, irrigation schemes and expert education and advice on sustainable methods of farming. Most of these initiatives will be supported by international aid.

What can be done to reduce the impact of climatic hazards?

Key ideas	**Content detail**
There are a variety of ways to manage or reduce the impact of climatic hazards.	• The extent to which climatic hazards can be predicted. • Management strategies to reduce the impact of climatic hazards.

Key questions

To what extent is it possible to predict climatic hazards?

Climatic hazards are among the most predictable of all natural hazards, especially those that develop slowly over days or weeks. Tropical storms are monitored by weather satellites, aircraft, radar, buoys moored at sea and radiosondes. Data are fed into computer models that forecast storm tracks and intensity. In the USA the National Hurricane Center is an agency devoted exclusively to monitoring and forecasting storms in the North Atlantic, Caribbean and eastern Pacific. It issues warnings to the media and on its own website of approaching hurricanes.

In the UK the Meteorological Office provides two types of weather warning: weather watch and severe weather warning. Weather warnings are broadcast when extreme weather is likely to disrupt transport, damage infrastructure and present a significant risk of death or injury. Generally the warnings are highly accurate up to 2 or 3 days in advance, though it is more difficult to forecast exactly which places will be hardest hit.

Forecasting tornadoes is more difficult. They are short-lived, affect only very small areas, and their formation is not fully understood. General alerts are issued when atmospheric conditions favour tornado development but it is impossible to say exactly where or when a tornado will strike. The highest level of alert is a tornado warning, i.e. a tornado is imminent and has been either sighted or indicated by radar. At this stage people are advised to find shelter — ideally in a storm cellar basement.

Droughts develop slowly, usually over many months. In western Europe, droughts are associated with anticyclones and the prolonged blocking of the prevailing westerly airflow. Today, general pressure patterns can be forecast with some accuracy up to a fortnight in advance. The development or continuance of drought is therefore predictable and provides vital information to farmers and water authorities managing public water supplies.

What strategies can be used to reduce the impact of climatic hazards?

Storm surge hazards can be reduced by building **'hard' engineering** structures such as cyclone shelters and levées. Ecological approaches that offer an alternative to 'hard' engineering, focus on the sustainable management of coastal ecosystems and environments.

Cyclone shelters, built on stilts, provide temporary refuges in low-lying deltaic environments. In the past 20 years they have saved thousands of lives in the Brahmaputra-Ganges delta in Bangladesh. **Basement shelters** provide similar protection against tornado hazards and are widely used in the US Midwest. Levées are flood embankments designed to contain storm surges and prevent flooding. Their effectiveness depends on their height and their maintenance. When cyclone Aila hit Bangladesh in 2009, it destroyed 1,700 km of levées. Many of these levées will never be rebuilt and what was productive farmland will be abandoned.

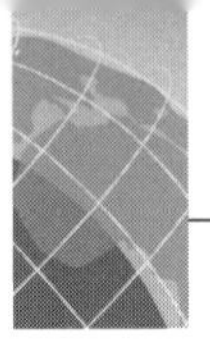

Drought hazards can be tackled by increasing water storage capacity (e.g. building more reservoirs), increasing water supply (e.g. drilling more boreholes and wells) and transferring water from areas of surplus to areas of deficit. The last involves the use of rivers for water transfer and the construction of aqueducts and pumping stations.

Ecological approaches to hazard mitigation work with natural ecosystems and in the long term are sustainable. Deltaic areas at risk from storm surges can be protected by dismantling levées and encouraging natural processes of sedimentation (through river flooding) that gradually raise land levels. On exposed tropical and sub-tropical coastlines the conservation of vegetation, especially mangrove forests, can greatly reduce the impact of storm surges. Flooding, caused by extreme rainfall, can be controlled by management that promotes afforestation in headwater catchments and protects wetlands. In large urban areas, extreme summer temperatures could be mitigated by creating more green spaces and planting more trees, thereby increasing evapotranspiration and cooling.

In what ways do human activities create climatic hazards?

Key ideas	Content detail
Human activities may impact on the global climate to create particular climatic hazards.	• The causes of global warming — increasing levels of greenhouse gases as a result of combustion of fossil fuels, deforestation etc. • The effects of global warming on climate change, sea level and biodiversity. • Global dimming — its causes and effects. • The causes, impacts and solutions to acid rain and photochemical smog.

Key questions

What are the causes of global warming?

The current rise in average global temperature is almost certainly due to human activities, which have significantly increased levels of atmospheric carbon dioxide and other greenhouse gases. These gases, which are largely transparent to incoming solar radiation, trap increasing amounts of long-wave radiation emitted by the Earth — a process called the **enhanced greenhouse effect** — and are responsible for the steady rise in global temperatures. Several human activities input greenhouse gases to the atmosphere. Most important is the burning of fossil fuels, which has increased massively in the past 50 years. Deforestation also releases large quantities of carbon dioxide that was previously stored as carbon in trees.

What are the effects of global warming?

The effects of global warming are wide-ranging and if unchecked will almost certainly be disastrous for the biosphere and for society. The most obvious effect is climate change. Climatic belts will shift polewards. Semi-arid regions like the Mediterranean, and the continental tropical and temperate grasslands will become drier, making farming unsustainable. As a result, millions of people could be displaced. Elsewhere, heatwaves will occur more frequently, bringing major health problems to large cities in mid-latitudes. In a warmer world, evaporation will increase and rainfall, violent storms and river flooding will be more common. The greater intensity of storms will accelerate coastal erosion and increase the coastal flood risk.

Rising temperatures are already causing ice sheets, glaciers and sea ice to melt at an alarming rate. As the ice retreats and exposes land and sea surfaces, **positive feedback** raises global temperatures even higher. The melting of land-based ice will raise sea levels by 2 m or more by the end of the century. In developed countries coastal defences will have to be strengthened, but in poor countries like Bangladesh, millions of people will become environmental refugees. Some island states such as the Maldives and Tuvalu, which are barely above sea level today, will be lost completely. Meanwhile, melting of the permafrost in the Arctic and sub-Arctic could release catastrophic amounts of carbon dioxide and greatly speed up the rate of warming and sea level rise. Global warming could drastically change the **thermohaline circulation** and the pattern of ocean currents, with potentially devastating impacts on regional climates.

Many plants and animals will be unable to adjust to changing climates, leading to mass extinctions and a loss of **biodiversity**. Species found in higher latitudes and in mountain habitats are most vulnerable. As the climate warms up they will literally have nowhere to go. Finally, in warmer and more humid conditions, many tropical diseases such as malaria and yellow fever are likely to diffuse polewards.

What are the causes and effects of global dimming?

Global dimming describes the gradual reduction in the amount of solar radiation reaching the Earth's surface in recent decades. It is caused by tiny airborne particles or **aerosols** released into the atmosphere by combustion, dust storms and volcanic eruptions. However, human activity through the burning of coal and, to a lesser degree, oil and gas is thought to be the main driver of global dimming.

The effects of global dimming could be both positive and negative. On the positive side it could reduce incoming solar radiation and therefore help to cool the planet. The negative effects include possible reductions in **photosynthesis**, plant growth and crop yields, especially in middle and higher latitudes. It is also suggested that global dimming could increase the incidence of drought, due to lower evaporation and disruption of the hydrological cycle.

What are the causes, effects and responses to acid rain?

Emissions of sulphur dioxide, nitrogen oxides and ammonia from coal-fired power stations, motor vehicles and intensive farming pollute the atmosphere. Mixed with water droplets in clouds, they eventually return to the surface as **acid rain**. The principal effects of acid rain are:

- the acidification of lakes, streams and the soil
- the destruction of forests and wildlife habitats
- the corrosion of buildings and stone monuments

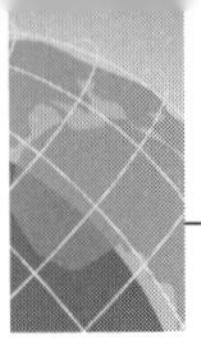

Acidic soils are **leached** of essential nutrients, and toxic metals such as aluminium get into food chains and destroy aquatic life. Toxic soils and dry acid deposition on foliage caused widespread destruction of coniferous forests in Europe in the 1970s and 1980s. Acid rain also adversely affects human health and contributes to respiratory diseases such as bronchitis and asthma.

In Europe there has been a concerted international response to the acid rain problem, and levels of acidity in rainfall have been significantly reduced. Binding emissions targets to be achieved by 2010 were set in the Göteborg Protocol (2001) for individual EU countries. As a result of this and earlier initiatives, emissions of sulphur dioxide and nitrogen oxides fell by almost three-quarters between 1980 and 2004. Liming of acidified lakes was adopted as a solution of last resort in Scandinavia, and in the short term also proved successful. Although the acid rain has been effectively tackled in Europe, it remains a serious environmental problem in rapidly industrialising countries such as China and India, which rely heavily on coal-fired power stations for electricity.

What are the causes, effects and responses to photochemical smog?

Photochemical smog is a modern form of air pollution found in large urban areas. Its main cause is emissions of pollutants such as nitrogen oxides and particulates from motor vehicle exhausts and industries. These pollutants react with sunlight to form photochemical smog, which is a major health hazard in some of the world's largest cities. Old people and children are particularly vulnerable; so too are individuals who already suffer from respiratory problems such as asthma, bronchitis and emphysema. Healthy people are aware of the smog because of the irritation it causes to eyes, noses and throats. The visible evidence of the smog is the brown haze that hangs over many cities during the summer months.

Ideal conditions for the development of photochemical smog include warm sunny climates and large concentrations of motor vehicles. Local geography may create temperature inversions which exacerbate the problem. Santiago de Chile, located in a basin and surrounded by high mountains, experiences frequent temperature inversions which trap the smog and produce very high levels of air pollution. Similarly, inversions develop in Los Angeles, where cold dense air from the Pacific Ocean moves onshore and is trapped against the high coastal ranges. Warmer air, displaced above the colder air, forms an inversion, trapping the pollutants from millions of cars. In California's warm sunny climate, the result is photochemical smog.

Santiago aims to reduce air pollution by encouraging less polluting forms of transport. The city's ancient bus diesel fleet will be replaced with clean technologies such as hybrid diesel-electric and compressed natural gas (CNG) vehicles. Road pricing is also being considered as a way of reducing traffic volumes in the city. Los Angeles has recently ordered a fleet of 260 CNG buses and is also extending its rapid transit metro system.

Economic issues

Population and resources

How and why do the population size and the rate of population growth vary over time and space?

Key ideas	Content detail
Population is dynamic and changes in response to the interaction of demographic, social, economic and political factors. The factors vary from place to place.	The study of how and why populations grow over time to illustrate: • global contrasts in population growth • how the rate of growth changes over time • the roles of natural increase and migration • how population change is influenced by a combination of demographic, social, economic and political factors • population growth related to over- and underpopulation

Key questions

What is the global pattern of population growth?

Current global population is around 6.7 billion, with a figure of 9.4 billion anticipated by 2050. The average annual growth rate is about 1.3%, resulting in some 80 million individuals being added each year. However, there are significant spatial variations from this average at the global scale.

The basic division in terms of population growth is between the developing countries with their high rate of growth (about 1.37% per annum) and the developed countries, where growth is much lower (about 0.22% per annum).

Within these two groupings there are regional contrasts in growth. Sub-Saharan Africa and the Middle East have higher rates of growth than Asia, Latin America and the Caribbean, where growth rates are declining. Contrasts in growth are less pronounced in the developed world. North America and the more developed parts of Oceania (e.g. Australia) have the highest rates of growth. The lowest rates of growth are in eastern Europe, Russia and the Baltic region.

How have rates of world population growth varied over time?

Around 10,000 years ago, humans were learning to cultivate crops and domesticate animals, a period known as the Neolithic Revolution. The estimated global population was about 5 million and was increasing at less than 0.1% per annum. Growth was very slow so that by about 2,000 years ago the global total was only 300 million. The total passed the 500 million mark around the middle of the seventeenth century but took until 1830 to double to 1 billion.

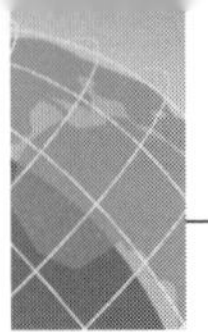

The twentieth century has witnessed unprecedented growth with a four-fold increase from 1.65 billion in 1900 to 6 billion in 2000. Globally the growth rate has been steadily declining since peaking at 2.2% p.a. in the early 1960s.

What role does natural increase play in global population change?

Natural change in a population is due to the interaction of births and deaths. When births exceed deaths, **natural increase** results; natural decrease occurs when the number of deaths exceeds the number of births.

At the global scale, the difference between births and deaths is responsible for population change. **Crude birth rates** (CBRs) and **crude death rates** (CDRs), known as **vital rates**, are widely used measures of population change. They are, however, of limited value as they are simple ratios of births/deaths to total population and take no account of age structures.

More accurate assessments of natural change rely on measures of fertility and mortality. The **total fertility rate** (TFR) is the average number of children that a woman gives birth to in her lifetime. **Age-specific mortality** relates the number of deaths in a population at a given age to the total number of people in the same age category. **Infant mortality** is the number of child deaths of those under 1 year old per 1,000 live births. **Life expectancy** at birth also gives an effective indication of mortality.

What role does migration play in population change?

Migration is the permanent or semi-permanent change of residence by an individual or group of people. **Net migration** is the balance of flows of people into and out of an area and can be either a loss or gain. The composition of these flows in terms of age, gender and socioeconomic status of the migrants, often has significant effects on the development and exploitation of resources.

Many migrations are **age selective** and influence natural population change. If an area receives large numbers of retired migrants, then natural decrease may occur. Natural decrease can also occur in regions experiencing substantial out-migration of young people. Locations receiving large net influxes of young people are likely to experience natural increase, leading to population growth.

How do demographic factors influence population change?

The key demographic factors that contribute to population change are **age–sex structure** and migration. Population structure varies considerably at different scales. At the global scale, developing countries have a relatively youthful population while developed countries have an **ageing** population. A shift of focus to the national scale reveals striking contrasts. Many developing countries have a high proportion of young adults or children, so death rates are low — even lower than in many developed countries. This is particularly the case in those countries currently undergoing rapid economic growth, such as Taiwan and Malaysia.

In the developed world, the high proportion of the middle-aged and elderly results in raised mortality rates. Some developed countries have even experienced natural decrease as fertility has fallen to very low levels. The ageing of these populations has implications for their development and exploitation of resources.

Given the high proportions of children and young people in developing countries, the next few decades are likely to see significant natural increase in these countries, as increasing numbers of young adults marry and have families of their own. This is known as **demographic momentum**. Even though fertility is declining in many developing countries, substantial increases in population are inevitable and will impact on the development and exploitation of resources.

At the regional and local scales, **migration** is often the driver of population change. Migration is a selective process influenced by factors such as age, gender, stage in family life cycle, education and employment. Young adults are often the most mobile group, with few social ties and ambitions to start independent lives. Many retired people, especially in rich countries, have freedom to choose where they live. As a result, migration flows of the over-60s have become an important part of the population geography of the developed world.

Young men make up the bulk of rural-urban migrants in many developing countries, although in some regions (e.g. South America) they are outnumbered by female migrants. Migration is largely driven by employment opportunities in the urban centres.

How do social factors influence population change?

Fertility is strongly influenced by social factors. **Cultural perceptions** and **interpretations of religious attitudes** towards family size, length of time between pregnancies and family planning techniques are important. It is, however, too simplistic to assume that the same religion exerts the same influence everywhere. Roman Catholicism is the majority religion in the Philippines and Italy. Yet in the Philippines, religion exerts a strong influence on government policy (e.g. abortion is banned and funding of family planning is limited). In contrast, religion has little influence in Italy. The outcome is a total fertility rate of 1.3 in Italy compared to 3 in the Philippines.

The **role and status of women** has an important influence on fertility. Many women in developing countries have little formal education and tend to marry at a young age, which means they start having children early. They often continue to be occupied in child-bearing and -rearing, which maintains their lower economic status. In many societies, a man's social status is enhanced by having large numbers of children. In the developed world the situation is reversed, with women achieving higher educational levels, delaying the age of marriage and pursuing formal careers outside the home.

How do economic factors influence population change?

There is a clear association between rising wealth and lower fertility levels. In large parts of the developing world, children make an important economic contribution to family income. This is impossible in developed countries, where children spend

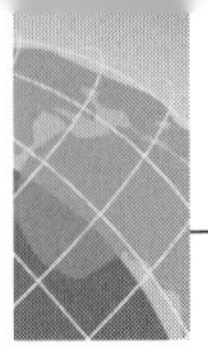

much longer in formal education. In the developing world, although each additional child represents another mouth to feed, children can help in the family business such as a farm or small workshop. Additionally, because poor countries are unable to provide state pensions or welfare payments, children are needed to look after elderly parents.

In the developed world, children often remain financially dependent on their parents until their early twenties. The opportunity costs incurred force many couples to delay having children. Some developed countries are concerned about falling fertility rates and have introduced fiscal incentives to encourage couples to have more children. Tax rules can be used to ease the economic cost of child-rearing, as in France and Italy.

There is a close link between wealth and mortality at both the individual and national levels. The relationship between poverty and infant mortality and life expectancy is strong. Poverty is associated with inadequate diets, sub-standard housing, contaminated drinking water and lower provision of healthcare. Average life expectancy is some 15 years more in the developed world than in the developing world.

How do political factors influence population change?

Governments of all types often have a direct influence on mortality, fertility and migration. Through their investment in public infrastructure, they aim to lower mortality and have helped improve the quality of life and life expectancy. They provide clean water, safe disposal of sewage and refuse, establish minimum housing standards and sponsor health services such as mass inoculations against disease.

Explicit policies aimed at influencing fertility fall into one of two groups: **pro-natalist** and **anti-natalist**. The former aim to increase fertility, the latter to reduce it. With the general decline in fertility in the developed world, several countries are now pursuing active pro-natalist policies. Various fiscal incentives such as tax breaks, cash payments, subsidised child care and paid maternity leave are examples of the measures used. For instance, in Russia, where demographic decline has been largely due to natural decrease, financial incentives worth around the equivalent of 2 years' average income are offered to women who have more than one child.

In contrast, many poorer countries pursue anti-natalist policies. The measures can be voluntary, with government focusing on education and publicity and increasing the availability of contraceptives. Countries faced with more serious demographic issues have adopted more stringent approaches. China's two- and then one-child policy (1979) has had periods of strict enforcement. One-third of all families in China are now single-child households. The latest forecasts suggest that China will achieve zero population growth by the mid-2030s.

Governments often control movements across their borders for demographic reasons. If a country has a labour shortage, in-migration of young adults can be encouraged. Australia in the 1950s and 1960s actively sought young immigrants to help their economy grow. At the same time the UK encouraged inward flows of people from new Commonwealth countries to supplement its workforce. The relaxing of border controls among member states of the EU has stimulated a large increase in international migration as migrants respond to changing economic conditions in both source and destination locations.

What are overpopulation and underpopulation?

Overpopulation is an excess of population compared to resources; **underpopulation** is when more people could be supported given the resources available. Between the two is **optimum population**, when a population maximises the ratio between population and resources at a sustainable level.

The ability to make resources available depends not only on where they are located but also on technology. For example, a valuable mineral resource like platinum can only be accessible if the technology exists to bring it to the surface.

Carrying capacity describes the relationship between the number of people and the resources that can be exploited sustainably. However, it is a controversial idea. The sustainable limit of resource use is difficult to quantify and can vary with changes in technology. For example, advances in agricultural production, perhaps through enhanced irrigation or crop breeding, can lead to higher yields, which in turn can support more people.

Too often ideas about over- and underpopulation are confused with high and low population density. High densities are not necessarily an indication of overpopulation. For example, Japan has a high population density (330 persons per km^2) but is able to support its 128 million people at one of the highest standards of living in the world. Eighty per cent of Japanese live in towns and cities and the service and manufacturing based economy procures resources worldwide.

Over- and underpopulation and carrying capacity are concepts that are most useful in the context of subsistence and semi-subsistence economies that rely on local resources such as soils, pasture and water.

How is population growth related to over- and underpopulation?

At its most simple, if population growth outstrips the growth of resources, then resources per person must decline. In these circumstances resources will tend to be over-exploited, degrade and a situation of overpopulation will prevail. At an extreme level of overpopulation, a **Malthusian** situation might exist in which mortality increases and numbers fall as food supplies are insufficient to support the population. The relationship between resources and population increase as suggested by the Malthus model is open to criticism. Population growth was assumed to follow a geometric rate (1, 2, 4, 8, 16...), while available resources increased arithmetically (1, 2, 3, 4, 5...). The past 200 years have witnessed economic growth, with resources expanding faster than population growth. **Marxist** analysis asserts that it is unjust social, economic and political systems that lead to poverty, not a lack of resources.

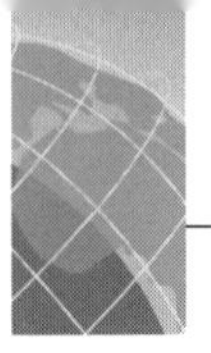

It is also worth noting that population growth rising above the optimum level does not automatically result in overpopulation. As population pressure rises, innovation can be stimulated. The maxim 'necessity is the mother of invention' summarises this view. **Ester Boserup**, an economist, argues that innovation is a crucial factor in the relationship between population growth and agricultural resources. Advances in agricultural technology in the second half of the twentieth century have led to increases in crop yields, especially in the developed world and South Asia. However, it is too simplistic to suggest a causal link between such advances in agriculture and the absence of population crashes. Social, economic and political changes are also important in this relationship. For example, improvements in the status of women, and in particular their access to education, are recognised as core changes in development. Increasing integration of economic systems, both nationally and internationally, can also result in greater **food security**.

Underpopulation describes the situation where there are too few people to exploit resources fully. For example, a region might possess substantial mineral resources for which there is insufficient labour to extract and process. Population growth would, therefore, allow development of the resource base and raise living standards.

How can resources be defined and classified?

Key ideas	Content detail
There are a variety of ways of defining and classifying resources, such as by source, by use and by how renewable they are.	The study of different types of resource to illustrate: • differences between renewable, non-renewable, flow and semi-renewable resources • how changes in technology and society can bring about changes in resource definition

Key questions

What is a resource?

There is no single agreed definition of a resource but in general it is something that satisfies human wants and needs. Resources can be identified at a variety of scales, from global down to an individual household.

A basic distinction can be drawn between **natural** and **human** resources. The former include substances, organisms and properties of the physical environment such as soil, water, fish and wind. These are valued because people perceive them to be useful in satisfying needs and wants. Human resources are features of the human population such as numbers of people, their abilities and skills. This is often referred to as **human capital**.

What is the difference between renewable and non-renewable resources?

A valuable way of investigating resources is to assess how **renewable** they are. A renewable resource is capable of regeneration within human timescales. Most biological organisms are renewable, such as many plants and animals, whether they are domesticated or wild. An important set of renewable resources are the **flow**

resources that have a permanent character and do not need regeneration. Wind, solar, tidal, geothermal, wave and most other water resources make up this group.

There is a group of **semi-renewable** resources that can regenerate over intermediate timescales of 100–1,000 years. Some plants such as the larger trees, some fish stocks and peat belong to this category. It is significant for the management of resources that both renewable and semi-renewable resources tend to be linked within ecosystems.

Non-renewable resources are finite on human timescales. Once used, these substances cannot be replaced. Fossil fuels and minerals are classic examples, although recycling of metals such as steel and aluminium can prolong their availability. Increasingly, landscape in the context of tourism and a resident population, is viewed as a non-renewable resource.

How can natural resources be classified?

Although the distinction between renewable and non-renewable is perhaps most useful, there are other ways of classifying resources. The **source** offers one alternative. For example, marine resources include fish and shellfish; soils produce crops and livestock; forests generate timber products; and geothermal energy comes from tectonically active regions.

Resource use is another way of classifying resources. For example, energy, food and building are three examples of resource groups. However, many resources have more than one use. For example, china clay can be used to make porcelain, paper, toothpaste and some medicines.

The **extent** of a resource in terms of its geographical distribution can be used as a basis for classification. Some resources like gold and copper are only found in specific geographical locations. Others are spatially dispersed such as forests, sand and gravel. These two categories are known as **point** and **diffuse** resources respectively. However, some resources cannot be easily allocated to one of these two categories. For example, oil deposits tend to be concentrated at particular locations, such as the Middle East, but within this region, individual oil fields can extend over considerable areas and distances. Thus when we consider the spatial extent of a resource, scale is an important factor. Changing the **scale of analysis** can alter the geographical pattern and the spatial extent of a resource.

How does technology influence resource definition?

With changes in technology, what was once considered to be of no value becomes valuable and what was once valuable ceases to have value. Perceptions of what constitutes a resource are largely influenced by the available technology.

Thousands of years ago flint was a valuable resource and was needed to generate fire and make tools. Today, flint has little use and therefore little value to most societies and is no longer perceived as a resource.

The discovery of uranium in the late eighteenth century led to its use in the glass and ceramic industries but it remained limited in value. Uranium's radioactive

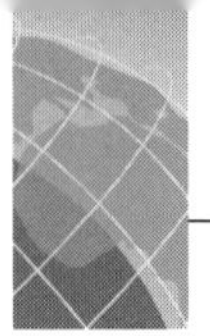

properties were recognised at the start of the twentieth century but it was only with the discovery of its fissile properties that it became a truly valuable resource. Today, uranium isotopes generate just under 20% of the world's electricity. This example of a relatively 'new' resource also highlights how resources may be used for both desirable (power generation) and undesirable (war) purposes.

Uranium is also interesting because of its geographical distribution. It is widely dispersed within the rocks of the upper crust but there are relatively few places where it exists in a sufficiently concentrated form to allow commercial mining. Just three countries — Canada, Australia and Kazakhstan — supply nearly 60% of the world's production.

What is the significance of society to resource definition?

Just as technology changes, so too do society and people's perceptions of resources. **Wilderness** was once viewed with suspicion and the prevailing attitude was to tame nature. Today wilderness areas are valued precisely because they are untamed; their landscape and ecosystems are conserved, often with the legal protection of National Park status. The growth in ecotourism has led to a reappraisal of some resources such as tropical rainforests, coral reefs and wetlands. In the past, traditional Balinese society disregarded beaches as they were unproductive in terms of food. But with the growth of international tourism, Bali has emerged as a major tourist destination and beaches have been reappraised and become resources.

Changing tastes within society can also bring about changes in resource definition. This can be due, for example, to consumer choice such as the preference for organic foods or the overexploitation of a resource. For instance, overfishing of cod has led to previously disregarded species like pollack becoming more popular.

What factors affect the supply and use of resources?

Key ideas	Content detail
The supply and use of resources is determined by a combination of physical and socioeconomic factors.	The study of different types of resource to illustrate: • how physical factors influence resource supply and use • how technology influences resource supply and use • how socioeconomic factors influence resource supply and use • how these factors change over time

Key questions

How do physical factors influence resource supply and use?

The geographical concentration of a resource influences its supply and use. The more a particular resource is concentrated in one location, the more likely it is to be exploited.

Geology is a significant factor in the supply and use of minerals, fossil fuels and water. Minerals close to the surface can be extracted cheaply by quarrying or strip mining. In Lancashire, the glass industry obtained high-quality sand by this means and in western Australia, iron ore is extracted in the same way. Minerals at greater depth are more difficult to extract. If the prevailing economic conditions (e.g. low prices) are unfavourable, extraction will not take place.

The availability of some renewable resources depends on **environmental conditions**. Wind, solar, wave and tidal energy are spatially variable and their development is controlled by environmental conditions. At the global scale, wind power has its greatest potential in the middle and high latitudes where westerly winds dominate. Water is another resource that is not ubiquitous, so its use varies geographically. This is illustrated by the global and regional distributions of irrigation schemes and hydroelectric power.

Soil fertility has a strong influence on agricultural systems. Lower soil fertility supports only extensive systems such as cattle ranching and hill sheep farming. But highly fertile soils promote market gardening enterprises in the Netherlands and wet rice padi in the Philippines and Indonesia.

How does technology influence resource supply and use?

Technology is probably the principal factor that influences the supply of, and demand for, natural resources. This starts with the ability to identify and locate resources accurately. Scientific discoveries and their applications, such as seismology, have allowed the mapping of many mineral deposits. More recently satellites have been used to survey a wide range of resources, such as minerals and agricultural crops.

Mining depths were restricted until efficient and reliable water drainage became available during the nineteenth century. The move of offshore oil drilling into deeper waters in the late twentieth century has relied on technological advances in construction and operation of the rigs. Meanwhile, advances in technology have allowed a higher proportion of oil to be extracted from a field. This is possible because engineers can control drilling in three dimensions and at considerable distance from rigs.

Technology also plays an important role in transporting resources from their source of origin to markets and in reducing transport costs. Advances in ship design and construction techniques, as well as navigation and control of individual vessels, have led to increasing capacities. Oil tankers of 500,000 tonnes are used to move crude oil between continents, while bulk cargoes such as mineral ores and grains are frequently transported by 150,000-tonne carriers.

Whether something becomes a resource also depends on technology. The ability to smelt various mineral ores has led, at different times, to changes in the dominance of individual metals. For example, bronze, an alloy of tin and copper, was the first metal to be smelted about 5,000 years ago but was then largely replaced by iron around 3,000 years ago when technology evolved further. It was not until the late nineteenth century that bauxite was first smelted by electrolysis to form aluminium. Since then bauxite has become a valuable resource.

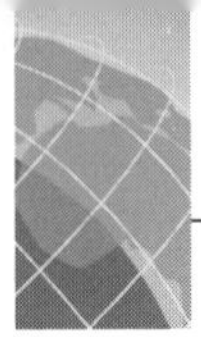

How do socioeconomic factors influence resource supply and use?

Of great significance to the supply and use of a resource are its economics. When the price of a resource exceeds its production costs, exploitation is likely to occur and vice versa. There are three economic terms used to describe resources in connection with their supply and use:

- **Resource base** refers to the entire presence of a substance that might become resources if they could be extracted.
- **Reserves**, sometimes known as recoverable reserves, are that part of the resource base that could be used given the prevailing economic, technological and social conditions. It does not mean that all such supplies are being actively exploited. For example, tin mining in Cornwall has experienced times of production and closure as the demand and price of tin has fluctuated. Cornish tin occurs in veins running through granite and is relatively expensive to mine compared to producers in China, Indonesia and Brazil.
- **Resources** are the proportion of the resource base that can be extracted given current economic, technological and social conditions.

Social, political and environmental factors influence attitudes towards the exploitation of resources. The extraction of minerals and fossil fuels is often controversial, for example the exploitation of oil reserves in the Alaskan tundra. The siting of wind turbines in upland and coastal locations and the construction of dams are also the subject of intense debate, locally, nationally and internationally.

Why does the demand for resources vary with time and location?

Key ideas	Content detail
• Different parts of the world have differing demands for resources. • These demands change through time and with development.	The study of different types of resource to illustrate: • how demand is influenced by population growth and standards of living • how patterns of resource demand in developed and developing countries and NICs change with population growth and rate of development

Key questions

What is the link between population size/growth and the demand for different resources over time?

As a population grows there is often a corresponding increase in the demand for resources. At the global and national scales, the relationship between expanding populations and resource demand has led to growing concerns about the sustainable use of a range of resources.

The relatively high rates of population growth experienced by most developing countries since the mid-twentieth century has triggered large increases in demand for land for cultivation and dwellings. This can be seen in the spread of agriculture into more marginal areas, overcultivation and overgrazing in many semi-arid environments, and the growth of urban slum housing areas on hazardous sites (e.g. steep slopes).

Equally, periods of population decline have lowered the demand for resources. In England in the fourteenth century, as a consequence of the Black Death, prices for agricultural products fell as fewer people were around to purchase wheat and wool.

What is the link between standard of living and the demand for different resources over time?

Demand for resources is not simply the result of numbers of people. The relationship between supply and demand is strongly affected by the ability of people to consume resources, which is closely linked to standards of living.

The contrast between average standards of living in the developing world and the developed world is clear and translates into contrasting demands for resources. For example, the high standard of living in developed countries relies on large quantities of cheap energy. Thus the demand for energy resources is much greater in the developed world.

The vast majority of the 6 million inhabitants of mid-eighteenth century England had very little disposable income and their primary concern was getting enough to eat. Today, the proportion of income taken up by 'essentials' such as food, clothing and housing costs, accounts for only one-third of average income. Clearly this has significant implications for resource demand.

China's economic growth and rising standards of living have stimulated unprecedented increases in demand for all types of resources. The county's booming manufacturing sector has created a demand for large quantities of raw materials and energy. Although a relatively resource-rich country, this demand has led to a global search for natural resources.

Like China, the Democratic Republic of Congo is also remarkably resource-rich. It has enormous reserves of copper, cobalt and manganese, timber, oil and HEP. However, demand for these resources mainly comes from overseas as the people of the DRC are among the poorest in the world.

It is interesting to note that even if a country is not rich in natural resources, so long as standards of living are high, resource demand will be high. Japan is such an example. Imported raw materials and energy sustain most of Japan's domestic demand, and about half of Japan's food comes from abroad. Since the middle of the nineteenth century, Japan has enjoyed one of the world's highest standards of living by relying on human capital (i.e. its highly educated and skilled workforce).

Increasingly, developed countries are seeing their domestic production of natural resources decline but are able to maintain high standards of living by sourcing materials worldwide.

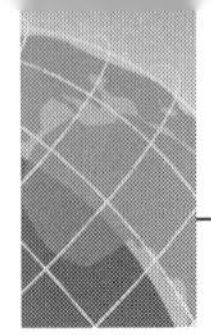

In what ways does human activity attempt to manage the demand and supply of resources and development?

Key ideas	Content detail
The demand for and supply of resources need to be planned and managed to achieve a sustainable system.	• The study of at least **two** different resources to illustrate contrasting types of planning and management strategies used to balance demand and supply. • At least **one** case study to illustrate attempts to make resource development sustainable.

Key questions

What is meant by 'sustainable resource development'?

In the late 1960s, the idea gained ground that common resources such as oceans and seas and the atmosphere were subject to increasing pressure from human activities. As a result these resources, whose use was unregulated, were steadily degraded. So long as individuals and groups benefited from these resources, exploitation continued even when the resources began to degrade and returns declined. The key principle is that although selfish behaviour may not be in an individual's long-term interest, neither is it in an individual's interest to be unselfish unless everyone is.

In 1972 a group called the Club of Rome generated computer models which forecast that continuing human use of many natural resources was likely to exceed the Earth's carrying capacity. The report's title, *Limits to Growth*, drew on Malthusian ideas to suggest the dangers of uncontrolled development.

Today, the most commonly quoted definition of sustainability is that given by the 1987 World Commission on Environment and Development Report, *Our Common Future*. It states that sustainable development is ***'development that meets the needs of the present without compromising the ability of future generations to meet their own needs'***. A key point in this definition is that a more efficient use of resources makes continued economic growth possible. However, many argue that greater emphasis should be given to ecological processes, population growth and the role of transnational corporations (TNCs) in the global economy. Increasingly, environmental assessments and environmental economics are being incorporated into development programmes in both the developed and developing worlds. It seems likely that the issue of sustainability will play a key role in the future debate on resource use and development.

How does human activity attempt to plan and manage the demand for resources?

Resource demand can be planned and managed at different scales. At one extreme there is the individual making personal decisions about their use of resources. Some people, for example, will decide not to own a car or might use energy-saving light bulbs in order to reduce their carbon footprint. At the other extreme, national governments and supra-national organisations such as the EU and the UN authorise research projects, develop policies and implement strategies designed to modify resource use. For example, the UK government, at both local and national levels, implements water conservation through regulations affecting toilet design and the volume of water used in flushing.

How does human activity attempt to plan and manage the supply of resources?

Human activities, at different scales, influence the supply of resources. Individuals can decide to use peat-free composts in their gardens and recycle metals, paper and glass. Likewise, governments can plan and manage the supply of resources.

Marine fishing is a classic example of the **tragedy of the commons** in operation. No one owns fish in the seas until they are caught, but all those who fish affect one another. Fishing in the seas around western Europe has been a contentious issue for centuries, but with the development of very large-scale fishing operations using sophisticated technology, fish stocks have been depleted faster than they can regenerate. The Common Fisheries Policy of the EU has attempted to regulate the fishing industry and achieve **environmental sustainability**. It has also tried to achieve **economic** and **social sustainability** for the industry and the communities that depend on it. Unfortunately, these aims have the potential for conflict — reducing the catch will reduce economic activity and incomes.

Each year limits are set for the **total allowable catch (TAC)** for each species of fish based on existing stocks and rates of regeneration. The TAC is then divided into quotas for each EU member state. In addition there are limits to the number of days a fishing boat may actually fish and to the size of the mesh of fishing nets. Political influences are significant and the scientific advice on levels of TACs is regularly exceeded. The policy remains controversial and so far has failed to prevent overfishing.

There is also potential for the sustainable management of **forestry** as wood is a renewable resource. Forestry contributes 8% of GDP to Finland's economy and timber and timber products account for 30% of exports. In total, 200,000 people are employed in Finland's forestry and wood products sectors. Meanwhile, the traditional role played by forests in Finnish society, recreation, hunting and food gathering, continues to be highly regarded among the population.

It is, therefore, crucial for Finland that its forests are managed to ensure environmental, economic and social sustainability. To achieve this, Finland has implemented a National Forest Programme (NFP). Only native species are grown, which encourages biodiversity and creates minimal disturbance to natural ecosystems. Regeneration of the forests is mainly by natural succession rather than by planting nursery-grown seedlings. The main period of logging is during the winter when the frozen ground minimises damage caused by heavy machinery. Along rivers, belts of uncut trees are left untouched in order to reduce pollution of the water from sediment carried in run-off. Environmental sustainability is also seen in the protection of the most sensitive wildlife sites. Tree felling uses the shortwood method — stripping the branches, which are left to decompose on the forest floor. The success of Finland's NFP is evident in the excess of tree growth over tree extraction that has occurred in the past 40 years.

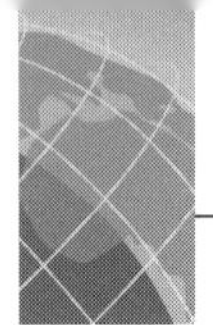

Globalisation

The meaning of 'globalisation' and why it is occurring

Key ideas	Content detail
• There are marked advantages for economic activity in working at a global rather than a local scale. • Globalisation is a complex process involving several interlinked components both economic and cultural. • A range of factors is responsible for globalisation. • Globalisation is a dynamic process and there is a variety of possible future trends.	• What is meant by the term 'globalisation'? • The process of globalisation is made up of several interlinked components. • What factors are responsible for globalisation? • What are the possible future trends?

Key questions

What is meant by the term 'globalisation'?

There is no simple definition of globalisation. As a formally recognised concept, it emerged in the 1960s with the idea of the **global village** which identified the growing **interconnection** and **interdependence** of people's lives. Globalisation compresses time and space so that the world is considered a 'smaller place'. This refers to the increasing speed and effectiveness of flows of capital, goods, services, people and ideas around the globe. Everywhere is looking and feeling more like everywhere else, an idea known as **placelessness**.

However, these ideas are not without their critics. Some argue that what is being experienced is simply an extension of **internationalisation**, the spread of economic activities across national boundaries. This is nothing new as throughout human history empires, tribes and states have moved out beyond their original home, often in search of additional resources. The process of industrialisation accelerated the trend and was frequently associated with **imperialism**. However, globalisation is recognised as being about the extent and quality of integration.

What are the components that are linked together by globalisation?

Economic activities are key components in globalisation. The production and selling of goods and services on a worldwide basis by transnational corporations affect the lives of billions of people. Whether it is the food people eat, the manufactured goods they use or the services they draw upon, increasingly there is a global element.

Globalisation has important social and cultural dimensions, which focus on the diffusion of western-style **consumerism** originating in the USA. Western media promote a lifestyle that is strongly influenced by corporations and conglomerates based in the USA and Europe. The creation of internationally recognised brands of goods and services encourages people (especially in the developing world) to aspire to lifestyles that differ from their local culture. However, the flow of culture is not one way, as illustrated by the spread of cuisine and music to the developed world from Asia, Africa and Latin America.

Closely associated with the cultural dimensions of globalisation is language. English has emerged as the predominant 'world' language, even though the great majority of English speakers have it as their second language.

Globalisation also has a political aspect. The growth in influence of supra-national organisations such as the UN, and trade blocs like the EU and the North Atlantic Treaty Organization (NATO), increase cooperation, trade and interdependence between countries. Influential **non-governmental organisations** (NGOs) such as Amnesty International and Greenpeace focus attention on international political issues. Global environmental conferences increase people's awareness of the interconnected nature of environmental systems and the need for global action if effective solutions to problems like climate change are to be found.

Demographics are another element of globalisation. The spread of different cultures around the globe has been facilitated by increased mobility. The movement of people across international borders, both for work and leisure, is greater than ever before.

What factors are responsible for globalisation?

Just as there are different components contributing to globalisation, there are different factors responsible for it. Economic integration has received a significant boost from international arrangements that have liberalised trade and finance. The World Trade Organization (WTO) has been instrumental in the massive growth of international trade in the past 15 years or so. The relaxation of controls on international capital movements has stimulated flows of money and the growth of financial markets.

Meanwhile the fall of the Berlin Wall in 1989 marked the start of growth in the economies of eastern Europe and the former Soviet Union. Countries such as Hungary and Poland have been drawn into the global economic system. Cuba, once a close ally of the Soviet Union, is now an important tourist destination for visitors from North America and Europe.

The emergence of the **New International Division of Labour** (NIDL) is both a cause and effect of globalisation. It refers to the migration of substantial parts of the secondary sector from the developed world to **emerging economies** and developing countries. **Supply chains** for many products now stretch across international borders and continents. Low labour costs encourage many manufacturing companies to move from the developed world to the developing world. This trend is most apparent among labour-intensive industries such as textiles, clothing and domestic electronic goods. As skill levels and wages have risen in **newly industrialising countries** (NICs) like Taiwan and South Korea, production has shifted to lower-cost countries such as Cambodia and Vietnam.

Technological improvements in **transport** have facilitated the international movement of goods and the emergence of the NIDL. Bulk carriers, transporting raw

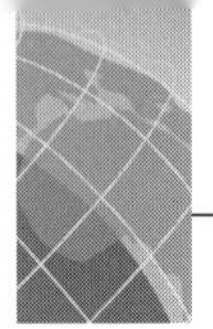

materials such as oil, mineral ores and grains, carry such large cargo in one trip (up to c350,000 tonnes) that the cost per tonne/km is significantly reduced. This is known as **economies of scale**.

Even more significant is **containerisation**. By using containers to move goods, transport costs for general cargoes have been greatly reduced. In addition, the improvements in packaging allow virtually all semi-finished and complete goods, even if relatively fragile, to be moved safely with minimal damage en route.

Many TNCs from eastern Asia have opened manufacturing plants in North America and Europe to be near to their markets. **Foreign direct investment (FDI)** of this type allows companies to be in close contact with markets and avoids trade barriers such as tariffs and quotas.

Technological advances in voice and data communications such as satellites and computers have allowed services to be provided from places that are distant to their point of consumption. This **offshoring** of service activities, such as accounting and call centres, reduces costs for companies based in developed countries and provides jobs in places like Bangalore, India.

What are the possible future trends?

Globalisation will continue to expand, although its progress will be influenced by global economic cycles. Events such as natural disasters and political tensions affect economic, social and environmental systems, which feedback into the globalisation process. There is also considerable opposition to globalisation from around the world (e.g. demonstrations at the G8). Protest groups are part of the rise of a **global civil society**, a concept describing transnational social concern and activity as people become more connected with each other.

Some commentators believe that the rise of TNCs means that nation states will have less power to influence global and national economic systems. TNCs tend to operate with little regard to national identity and look for business opportunities anywhere in the world. National governments find it increasingly difficult to manage their economies when large-scale movements of money are controlled by international banks and investment funds.

The growth of **sovereign wealth funds** is another recent development in the globalisation process. These are state-owned investment funds looking for business opportunities around the world and not just within the borders of their own country. Many involve monies generated from oil and gas production with the three largest funds held by the United Arab Emirates, Saudi Arabia and Norway. China has the next two largest funds, made up of money from its large balance of trade surplus. Concerns have been expressed regarding the wisdom of relying on too much overseas investment and the consequent loss of domestic control over economic decisions.

The spread of western ideas regarding material possessions and the **secularisation** of society is being challenged by some religious groups. This can have a violent expression, as seen in types of religious **fundamentalism**.

What are the issues associated with globalisation?

Key ideas	Content detail
• The globalisation of economic activity has a variety of impacts, environmental, economic, social and political. • These impacts may bring advantages or disadvantages to various areas and/or groups of people.	• A study of the impacts of globalisation on a developed country. • A study of the impacts of globalisation on an NIC. • A study of whether globalisation is increasing or narrowing the 'development gap'. Included in this is statistical analysis.

Key questions

What are the impacts of globalisation on a developed country?

Recent globalisation has brought both advantages and disadvantages for developed countries. **Economic restructuring** has seen a huge expansion of the **tertiary** sector at the expense of the **primary** and **secondary** sectors as the comparative advantages of developed countries in primary and secondary activities declined. Inevitably, this led to mine and factory closures and job losses, with traditional areas of economic activity hardest hit. Mining (e.g. coal), iron and steel, ship building, chemicals and textiles declined in both relative and absolute terms. Most often such **deindustrialisation** was geographically concentrated in certain regions. Worst affected were regions such as northeast England, northeast France and the Great Lakes of the USA that specialised in a narrow range of heavy industries. Today, many manufactured goods are now imported to developed countries from newly industrialising countries (NICs) and developing countries.

More recently, tertiary activities in developed countries have been affected by globalisation. Many back-office clerical jobs and call centres have relocated to developing countries such as India, where costs are much lower. Such offshoring has benefits for developed countries, allowing them to focus on higher skilled activities such as research, product design and development and marketing.

Globalisation has seen investment flow into developed countries in the form of foreign direct investment (FDI). TNCs see opportunities in these countries in all areas of the economy. New manufacturing plants built by TNCS in the developed world (**transplants**) often incorporate the latest technology and are highly efficient. Toyota, Honda and Nissan operate factories in both Europe and the USA, which are among the most productive in the world. As a direct result of FDI, developed countries also benefit from wealth creation through jobs, taxation and export earnings. However, there are disbenefits. Foreign firms do not have the same quality of allegiance to a country as indigenous ones. They often set up **branch plants** that in times of economic downturn are vulnerable to reductions in capacity or closure.

Globalisation has implications for the **physical environment** in developed countries. As many of the traditional manufacturing industries closed, substantial tracts of urban land were left derelict. Former industrial sites often present problems

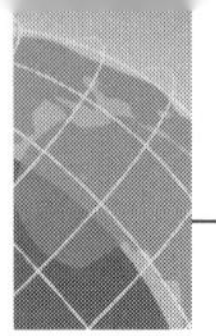

of visual pollution and contamination by toxins such as mercury and cadmium. Extensive reclamation of derelict industrial sites has been completed in places such as Swansea, Cleveland (Ohio, USA) and Bochum in the Ruhr, Germany. Waterfront sites have attracted redevelopment where docks and warehousing have closed (e.g. Newcastle-upon-Tyne and Hamburg, Germany).

Increased international mobility of workers has brought economic benefits through lower labour costs. Migrants also offer skills in areas where shortages exist. Most economic migrants are of working age and so contribute to wealth creation in an economy. However, in some parts of the UK, large influxes of migrants have placed considerable strain on local services. This has mainly been the case in rural areas (e.g. East Anglia) where migrants find temporary work in fruit and vegetable harvesting.

What are the impacts of globalisation on an NIC?

The economic growth experienced by NICs is causally linked to globalisation. Many current NICs stimulated economic development through **import-substitution**. Later they integrated their economies regionally and globally through international trade. Governments frequently took a lead role in promoting and planning such changes.

Rates of economic growth recorded by NICs have often been spectacular. During the past 20 years China has regularly achieved annual growth rates of 10% or more, while Taiwan and South Korea averaged between 8% and 9%. Such **wealth creation** has led to significant reductions in poverty for millions of people and provided governments with the resources to invest in infrastructure, health and education projects. However, the benefits of growth accruing from globalisation have not always been shared equally across society. In China, workers in agriculture have lagged behind those in manufacturing and services. More geographically remote regions have tended not to receive FDI.

The growth of the manufacturing sector has gone hand in hand with **urban growth**. Guangdong province in southern China, an area of rapid economic growth linked to FDI and export-based manufacturing, is China's most urbanised province. In South Korea, Seoul's population has risen from 8.5 million in 1980 to over 10 million today. Urbanisation and industrial growth have led to high levels of environmental pollution. Air, water, noise and land pollution often exceed World Health Organization safe levels. Clean-up operations have been implemented in the worst affected areas but with varying degrees of success.

NICs are not immune to **economic problems** related to globalisation. Because they are integrated with the global economic system, they experience periods of recession as well as growth. In several NICs wage rates have been rising, which has reduced the country's competitiveness and led to some industries moving on to other countries with lower production costs. For example, some TNCs have moved assembly production of goods such as televisions from Taiwan and South Korea to Vietnam and Thailand.

Politically many NICs have become more democratic as they have been absorbed into the global economic system. They have also demanded greater participation in world bodies that make decisions about trade (WTO, G20), where membership may be conditional upon recognising human rights and establishing democracy. Inevitably this brings tensions, perhaps most strongly evident in China. There is also tension between some NICs and the USA and EU as a large trade imbalance has built up in favour of the NICs. The activities of sovereign funds also cause concern in the developed world.

How has globalisation affected the 'development gap' between the developed world and the developing world?

Some developing countries, especially in Asia and Latin America, have participated actively in globalisation and increased their per capita incomes and living standards. However, Africa, and in particular sub-Saharan Africa, has fallen further behind, increasing the **development gap**. Today, 20 countries in sub-Saharan Africa have lower incomes per capita in real terms than they did in the 1980s.

Although some developing countries appear to have made economic progress, too often this is due to rising commodity prices. When prices fall, the economic effects can be catastrophic, especially where there is reliance on a narrow range of exports. Uneven and unequal development remains a characteristic of the global economic system. Competition for global capital is fierce, affecting both the developing world and the developed world. Globalisation seems likely to continue to produce winners and losers.

What are transnational corporations and what is their contribution to the countries in which they operate?

Key ideas	Content detail
Transnational corporations (TNCs) may create both positive and negative impacts on an area.	• What is meant by a TNC? • How have TNCs developed over time? • How do TNCs vary in their spatial and organisational structures? • What are the disadvantages and advantages of TNCs to countries at different levels of development?

Key questions

What is a TNC?

TNCs are large firms with production and marketing operations in at least one other country in addition to their home country. They are often **footloose** and have the economic power and resources to transfer operations quickly between international locations. They own and control overseas activities either directly or via joint ventures, licensing or franchising. Many manufacturing TNCs outsource production to a network of international sub-contracting firms.

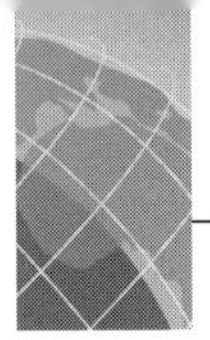

How have TNCs developed over time?

Since the sixteenth century large companies have sustained global operations. However, it is only since the mid-twentieth century that TNCs have emerged as a dominant force in globalisation.

Some TNCs arose from colonial links established by European powers during the nineteenth century. By 1939, USA-based TNCs were prominent overseas investors and their role has continued to expand. In the second half of the twentieth century, TNCs based in the USA, western Europe and Japan became major international players. Most recently, TNCs based in the NICs and BRIC group (Brazil, Russia, India and China) have emerged as competitors in the global market place.

How do TNCs vary in their spatial and organisational structures?

TNCs show considerable variation in size and internal organisation. Typically they have three components: a **head office**, **research and development (R&D)** and **production** locations, sometimes referred to as **branch plants**. The HQ is usually in the home country, generally in a **core region**. R&D is often located either with the HQ or close by. Although some older production plants remain in the country of origin, many have been relocated overseas. Some of the larger TNCs have delegated a degree of control and R&D to major regional sub-divisions (e.g. Ford's HQ is in Detroit and it has a European division in Cologne). The Swedish furniture manufacturer IKEA has its financial control and strategic decision-making HQ in Leiden, Netherlands. Its original location Älmhult, Sweden, retains the primary design function.

Many TNCs have used a **Fordist** approach to their manufacturing. This originated with Ford Motor's revolutionary assembly line in the early twentieth century. Building vehicles was broken down into simple activities on assembly lines leading to **mass production**. **Unit costs** were slashed and allowed for a fall in the real cost of many products, initiating the era of **mass consumption**.

Flexible production replaced mass production and assembly lines in some manufacturing industries in the last quarter of the twentieth century. With flexible production, cars can vary in the precise specifications such as engine type and capacity and internal finishing. Such a system requires more highly skilled workers and the availability of 'intelligent' computer-controlled machinery.

What are the disadvantages and advantages of TNCs to countries at different stages of economic development?

The country of origin of a TNC benefits from **wealth creation** within the country through wages, taxes and overseas earnings. The TNC stimulates the demand for skilled labour in management and research and promotes the profile of that country internationally. The disadvantages of TNCs include the loss of manufacturing to overseas competitors and the resulting negative effect on the country's trade balance.

Developing countries gain economic benefits from FDI by TNCs in wages and taxes. TNCs often pay higher wages than local firms and develop new skills among the workforce. Having attracted investment, a major TNC may encourage further investments from other companies. Exports generated by TNCs make a positive contribution to the balance of payments, and tourist-sector TNCs help to secure 'hard' currency.

The downside is that TNCs can easily transfer production elsewhere. In times of economic recession, remote branch plants are the first to close. **Political instability** may also encourage a TNC to transfer production or curtail future investment. **Leakage** of funds occurs when TNC profits are returned to the country of origin. The constant drive to reduce costs often comes at the expense of wages and working conditions. It can also mean a disregard for pollution controls or governments failing to enforce existing controls.

How far do international trade and aid influence global patterns of production?

Key ideas	Content detail
• Trade supports and hinders the broader balance of the world's patterns of production. • Aid supports and hinders the broader balance of the world's patterns of production.	• What is the pattern of international trade? • What has been the impact of international trade on developed countries? • What has been the impact of international trade on NICs? • What has been the impact of international trade on developing countries? • What is the role of international trade negotiations and agreements? • What are the different types of aid? • What are the advantages and disadvantages of aid for both donor and recipient countries?

Key questions

What is the pattern of international trade?

Substantial increases in world trade both in goods (merchandise) and services have occurred in the past 20 years. Most international trade is **intra-regional**, that is it takes place within one or other of the major world regions, such as Europe or Asia and Oceania. The greatest **inter-regional** flows are among Europe, North America and Asia, which result in developed countries dominating trade. Since 1980, the world's poorest countries have suffered a decline in their share of world trade.

Trade in services has grown rapidly over the past decade, almost exclusively between developed countries. However, China and India figure importantly and many developing countries receive large numbers of international tourists. The top 500 TNCs account for 70% of world trade.

What has been the impact of international trade on developed countries?

Most developed countries have increased the value of their international trade over the past few decades. Both merchandise (goods) and services are being traded globally on an unprecedented scale. Such increases represent both cause and effect of sustained international economic growth. Developed countries have gained

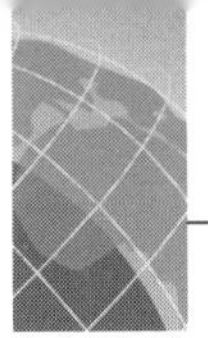

most from international trade. The USA dominates trade statistics both as exporter and importer. It is the second largest exporter of merchandise but also the largest importer. However, US merchandise imports significantly outstrip exports in value, so a substantial trade deficit has developed. While some are worried by this situation, others argue that the enormous US economy can cope. Germany, the world's second largest exporter and importer of merchandise, has a positive trade balance.

The UK ranks about seventh for merchandise exports and fifth for imports, which creates a significant trade deficit. The main area of trade is in manufactured goods but the recent growth in **commercial services** such as finance and insurance, has helped offset the negative trade balance in merchandise. Indeed, there is a trade surplus in services. Sixty per cent of the UK's imports and exports by value are with the EU. Two-thirds of the UK's trade is with just 13 countries and of these only one, China, is not a developed country.

What has been the impact of international trade on NICs?

NICs have also benefited greatly from the expansion of global trade. Compared to developed countries, their pattern of trade relies more on primary products and manufactured goods and much less on services.

Brazil, for example, currently runs a trade surplus, with agricultural products (soya, sugar, coffee, meat) and fuels and mining products (oil, iron ore, bauxite) making up half of its exports by value. Only a small proportion of trade comes from services, leaving manufactured goods to account for the rest. Brazil's manufacturing sector is diverse and includes some major TNCs such as Toyota and Ford.

The main destinations for Brazilian primary products are Europe and North America, while its manufactured goods are mainly exported to other countries in Latin America.

Undoubtedly Brazil gains from its trade. It is the world's tenth largest economy and has experienced high growth rates in recent years. It is not, however, immune to global economic cycles and within Brazil there are significant inequalities in living standards.

What has been the impact of international trade on developing countries?

On the whole, developing countries have not benefited from the expansion of global trade as much as developed countries and NICs. Their trade is dominated by the export of **primary products** and the import of manufactured goods. The former have little added value and so earn less for the country.

An additional problem facing many developing countries is that their exports often comprise a narrow range of commodities, especially agricultural products, fuels and mineral ores. Furthermore, prices for these products are highly **volatile** on world markets. This lack of stability in income makes long-term planning for development difficult. Many developing countries also struggle to gain advantages from international trade because prices for primary products have not kept pace with those for manufactured goods.

Many organisations support the idea of **fair trade** to ensure that a higher proportion of the retail price of a commodity goes to producers. Many foods are now sold as fair-trade items, including tea, coffee, sugar, cocoa and bananas. Although the fair-trade movement began as a charity, several supermarket chains have taken up the cause and stock fairly traded products on their shelves.

What is the role of international trade negotiations and agreements?

Trading rules are overseen at the global scale by the **World Trade Organization** (WTO). It evolved out of an earlier, less comprehensive agreement, the **General Agreement on Tariffs and Trade** (GATT). The idea driving the WTO is the liberalisation of trade. This is done by removing trade restrictions imposed by individual countries and groups of countries. Since 1945, average tariffs (i.e. charges on goods or services entering a country) have declined by about 90%. However, **protectionist** policies that shield a country's industries against competing imports still operate, sometimes openly but also in subtle ways. For example, some governments heavily subsidise their own export industries, undercutting the prices of their competitors. The WTO also recognises that simply opening up all economies to **free trade** can be detrimental for some countries, especially in the developing world. Negotiation is a key component of the WTO's operations.

There are several regional groupings of countries dedicated to their members' trading interests. The largest and most integrated **trade bloc** is the EU, which currently has 27 member states. Within the EU there is free movement of goods, services, capital and people. However, goods imported from outside the EU are subject to a charge (common external tariff). Meanwhile some EU industries such as agriculture and coal are subsidised to protect them from imports. Other trade blocs include the North American Free Trade Agreement (NAFTA), the Association of South East Asian Nations (ASEAN) and the Mercado Común del Sur (MERCOSUR) operating in South America.

What are the different types of aid?

Foreign aid consists of transfers of money, technology, expertise and training from rich to poor countries. In 1970 the UN stipulated that developed countries should aim to give a minimum of 0.7% of GNP in foreign aid. Only a handful of countries have achieved this level, all from northern Europe, with Norway leading the way. The USA is the largest donor in absolute terms (US$20 billion in 2008), though this represents just 0.16% of its GDP. The UK gave 0.36%.

There are two types of aid: **bilateral aid** and **multilateral aid**. The former is aid going directly from one country to another. Multilateral aid flows through an organisation such as the EU or the World Bank. In 1944 the **World Bank** and the **International Monetary Fund (IMF)** were set up. The World Bank lends monies

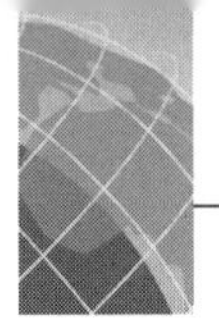

to developing countries for development projects. But increasingly it has moved into areas of policy and planning and has become a powerful global force on development projects. The IMF is more concerned with the health of the international monetary system and tries to ensure financial stability.

Non-governmental aid comes from organisations like Oxfam and Christian Aid. They are funded by private donations and they operate directly with partner groups in the recipient countries.

It is also useful to recognise the difference between **short-term emergency aid** such as relief given in the aftermath of a natural disaster (e.g. earthquake) and **long-term aid**. The latter focuses on projects such as infrastructural schemes (e.g. dams and roads) and developing human capital by training (e.g. teachers, nurses).

What are the advantages and disadvantages of aid for donor and recipient countries?

The advantages of emergency aid (disaster relief) to those receiving it are obvious — saving lives. Food, water, clothing, shelter and medical care represent a humanitarian response to another person's desperate need. NGOs are particularly valuable in such situations as they tend to have established relationships with local groups and can send the appropriate aid to those who need it most.

The issue of longer-term aid, particularly when it comes from governments, is more problematic. Aid is increasingly '**tied**' so that strict conditions are attached to its transfer. This can include reducing public spending, which results in cutbacks to health and educational programmes. In some aid projects, most benefits flow back to the donor country. Money for a dam project might be given as long as the HEP turbines are bought from a firm in the donor country. Aid can lead to the recipient country becoming **dependent** on outside help so that a state of underdevelopment persists.

Longer-term **food aid** can lower local food prices and affect local production and disrupt food-marketing systems. It can change local eating habits and lead to the neglect of plans to increase food self-sufficiency.

There has been a reappraisal of the nature of development projects. Rather than a **top-down** approach, where donors and central government determine large-scale projects, a small-scale **bottom-up** approach is preferred. This gives local people and organisations more say in what and how projects should be pursued. They become stakeholders and the direct beneficiaries of aid.

How can governments evaluate and manage the impact of globalisation?

Key ideas	Content detail
Governments vary in their responses to the impacts of globalisation and are increasingly looking to reduce the harmful impacts.	The study of one country to illustrate how the impacts of globalisation on its economy and society are being managed.

Key questions

How are the economic and societal impacts of globalisation managed at a national scale?

Globalisation has its advocates and its critics. It creates winners and losers: most developed countries and NICs receive a net benefit. The benefits are less clear for developing countries. In a general sense, no matter where a country lies along the development continuum, the rich gain more than the poor.

Although most developed countries claim to favour free trade, they often erect trade barriers to protect some sectors of their economies. Agriculture in the EU countries is given generous subsidies and price support. Some manufacturers receive government help via preferential loans and financial support for research and development. Governments also support services such as banks because of their vital importance to national economies.

Globalisation has stimulated international migration, raising concerns in receiving countries. The USA has strengthened its border with Mexico. The UK's new immigration policy includes a points-based system to assess the desirability of admitting economic migrants from outside the EU.

Developed countries are anxious to attract FDI but at the same time try to exercise control over it. For example, Japanese vehicle manufacturers in the UK have been 'urged' to source two-thirds of their components within the UK. Government grants encourage foreign TNCs to locate their factories in areas of high unemployment.

Some developing countries have been active in their involvement in the activities of TNCs. Bolivia, for example, nationalised its gas and oil industry in 2006, which meant the TNC energy companies had to renegotiate their terms of business. Generally, however, developing countries have little power in the face of the global strength of most TNCs.

Development and inequalities

In what ways do countries vary in their levels of economic development and quality of life?

Key ideas	Content detail
• Levels of economic development and quality of life can be assessed both quantitatively and qualitatively. • Levels of economic development and quality of life vary geographically at a global scale.	The study of global patterns of economic development and quality of life to illustrate: • different ways of measuring the level of development and quality of life (both quantitative and qualitative) • the contrast in the level of development and the quality of life between developing countries, NICs and developed countries (with the aid of statistical analysis and case studies)

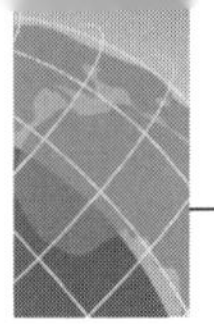

Key questions

What is meant by development?

Economic development relies on creating wealth by using natural and human resources to produce commodities (primary products), manufactured goods and services. The more value added to goods and services, generally the more secure economic development is.

Development involves more than just **economic growth**. There are **social**, **cultural**, **political** and **environmental strands** and all are inter-related. For example, the provision of clean water has an environmental element as water must be abstracted, stored and treated before use. Clean water is fundamental to improving health — a social strand. The quality of education, an element with both social and cultural aspects, influences attitudes towards clean water provision and its use. Politics can determine how the water is provided and how it is accessed by people. Improving economic conditions can lead to a rise in **standard of living**, the material nature of development. However, this does not necessarily translate to an improvement in **quality of life**, as indicated by factors such as overcrowded living conditions, quality of air and length of working day/week. Increasingly it is the **sustainability** of development that is crucial.

What quantitative measurements can be used to measure levels of development?

The most common measure is **gross domestic product (GDP) per capita**. This is the total market monetary value of all goods and services produced within a national economy in a given year divided by its population. It is more useful when adjusted to allow for variations in cost of living known as **purchasing power parity (PPP)**. An alternative measure is **gross national income (GNI) per capita**. This is the dollar value of a country's annual income divided by its population and includes income earned overseas. Both measures are deficient. For example, they take no account of non-monetary activities in an economy (such as subsistence farming and people making their own clothes), nor of the **informal sector** (such as street hawking).

Several non-monetary quantitative measures can be used to assess development. **Energy consumption per capita** is valuable because all sectors of an economy rely on energy, including essential services that influence the quality of life. No hospital can function without energy. Other widely used non-monetary measures of development are:

- infant mortality
- life expectancy
- literacy rates
- persons per doctor

The UN's **Human Development Index (HDI)** for countries is a composite measure based on life expectancy, adult literacy and school enrolment and GDP per capita (PPP in US$). The index ranges from 1.0 (most developed) to 0.0 (least developed).

Generally an index of 0.8+ indicates a high level of development, between 0.5 and 0.79 a medium level and less than 0.49 the least developed.

What qualitative measurements can be used to assess levels of development?

Quantitative data alone do not convey an accurate picture of what development means for individuals. The long-term toil for improvement in living conditions and the short-term struggle for survival in the aftermath of a natural disaster are not revealed in bland statistics. The media offer a variety of qualitative measures of development. Images and words are readily available through television, films, newspapers, magazines and novels. These sources are more likely to provide a meaningful insight into the social, cultural and psychological conditions faced by people in their daily lives.

What are the global contrasts in economic development and quality of life?

However measured, there are great contrasts in levels of development and quality of life at the global scale. Although development is a **continuum**, it is often useful to group countries according to levels of development. The World Bank uses GNI per capita per year to identify three groups: **low income** (less than US$935), **medium income** (US$936–11,456), and **high income** (more than US$11,456). The 49 low-income countries account for 20% of the world's population but only 1.5% of the world's annual income. The high-income group accounts for 16% of the world's population but 74% of its income.

A simple two-fold division into the **developed world** and the **developing world** is popular. A more detailed look at the patterns reveals a group of **least developed countries (LDCs)**, most of which are located in sub-Saharan Africa. Another group are the **newly industrialising countries (NICs)** — countries that have been developing through industrialisation over the past 40 years, such as Taiwan and South Korea. Some of these now have living standards equal to those found in the developed world. There is also an emerging group of **recently industrialising countries (RICs)** dominated by China and India.

Countries possessing vast oil reserves such as Saudi Arabia and the UAE are difficult to place. Their economic wealth puts them in the UN's 'high income' category, yet many societal features (e.g. status of women) would put them much lower on the development scale. The **former communist** countries of eastern Europe are also difficult to place as their economies and societies are in transition from a centrally planned system to a free-market economy.

Both economic status and quality of life can change quite rapidly. Zimbabwe was once a relatively prosperous country in southern Africa, but a combination of severe drought and political mismanagement has led to years of negative change in GDP, hyperinflation, food shortages and poverty.

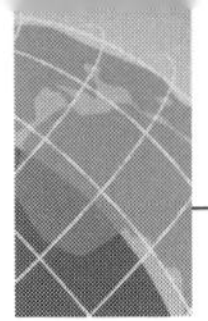

Why do levels of economic development vary and how can they lead to inequalities?

Key ideas	Content detail
Various factors influence the rate and level of development and this in turn may increase or decrease economic and social inequalities.	The study of the relative level of development of countries to illustrate: • the factors (physical, economic, social, political and historical) that influence the relative level of a country • how economic development can increase or decrease various inequalities between countries and within one named country

Key questions

How do physical factors influence development levels?

It is tempting to assume that a large natural resource base is a prerequisite for development. Some developed countries do have vast natural resources, such as the USA and Australia, but others like the Netherlands and Switzerland do not.

Climate has a role in supporting economic development. For example, it has a direct impact on agriculture and tourism. A range of climates within a country supports a variety of ecosystems that can be used to generate economic activity. Italy and France extend across more than one type of climate, allowing production from a range of agricultural enterprises and different types of tourism. In contrast, Chile and India also have a wide range of climates and ecosystems and yet have not achieved the same level of development. Harsh climates do not necessarily mean a lack of development as indicated by Canada and Finland, two of the world's richest nations.

It might be assumed that landlocked countries would not be as well developed as those who have ready access to trade via a coastline. Certainly many landlocked countries in Africa have struggled to make secure progress. Others such as Austria and Switzerland enjoy high levels of development.

How do economic factors influence development levels?

There is a very close link between economic factors and development. Countries towards the upper end of the spectrum, such as Denmark and Switzerland, have large and successful service sectors. At the lower end, countries often rely heavily on the primary sector, especially mineral extraction and agriculture (e.g. Botswana, Cambodia). As less value is added to primary products, less wealth is created.

Receipt of foreign direct investment (FDI) stimulates an economy and is closely associated with a country's involvement in international trade. Africa receives only a tiny proportion of global FDI (3.4%), whereas the EU gets nearly half (46%). However, it is difficult to establish cause and effect between trade and development as the two go hand in hand.

Efficient infrastructure such as power distribution grids, airports, road and railway systems and telecommunication networks is essential for development. An effective financial system is important to support the flow of capital within an economy.

How do social factors influence development?

The **education and skills** of a population are a key influence on development. Economic progress requires a literate and numerate workforce. The **knowledge-based** economies of the developed world rely on highly trained workers. **Emerging economies** such as Taiwan and India have placed great emphasis on education and the development of **human capital**. Labour-intensive agriculture creates little demand for a highly educated workforce and creates insufficient capital to pay for a sophisticated education system.

High rates of population growth are associated with low levels of development, although they are not necessarily a causal agent. Indeed, a growing market should stimulate economic activity. It is the relationship between the rate of economic growth and population growth that is significant. When population increase outstrips economic growth, then per capita incomes fall and vice versa.

Healthcare is a social factor that is both a cause and effect of development. If people are ill then they miss education and work, which reduces their economic potential. Malaria not only kills about 1 million people each year, it also debilitates many millions more. The World Health Organization estimates that countries where malaria is endemic suffer an average 1.3% loss in economic growth. The impacts of disease and natural hazards are much worse in the developing world as the healthcare systems cannot cope with demand.

The role and **status of women** have a direct influence on development. Access to education varies and it is in the poorer countries that female enrolment in schools is at its lowest. The effect is to reduce the pool of national human capital and hamper development.

How do political factors influence development?

Good governance plays an essential part in advancing economic development. Attitudes towards investment in education, health, infrastructure and trade have a key role. The dramatic change in the Chinese government's economic policies after 1978 resulted in spectacular economic growth. This parallels what happened in Japan in the late nineteenth century when change in government policies catapulted the country into the dominant force in Pacific Asia by the early twentieth century.

Bad governance is a major hindrance to economic development. Corrupt governments hold back development in many countries in sub-Saharan Africa. Extreme authoritarian regimes can have a disastrous impact. In Cambodia towards the end of the twentieth century the fanatical Khmer Rouge murdered thousands of its most educated people and dispatched others to the countryside for hard labour and 're-education'. As a result the economy collapsed.

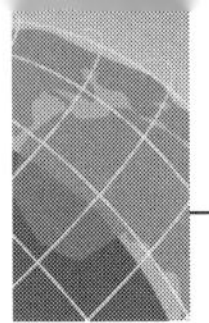

How do historical factors influence development?

Many developing countries are struggling with the legacy of **colonialism**, when their economies were run for the benefit of the colonising power. For example, their transport systems were often designed to do no more than move agricultural products and minerals to the coast for export. There was no attempt to develop integrated networks that would assist in economic development.

In Africa, new national boundaries were drawn up by colonial powers, which paid little heed to the realities of groupings on the ground. Thus, tribal and national differences were ignored. This has created a legacy of political tension and instability in countries such as Nigeria and Kenya.

How might economic development affect inequalities between countries?

Although the world is increasingly globalised, uneven and unequal development is the main characteristic of the world economic system. Competition for global capital for investment favours developed countries and spurs wealth creation. **Cumulative causation** is a process that reinforces some **initial advantage** such as a new resource or social change, which releases human capital. Growth stimulates growth so that successful countries maintain their position at the upper end of the development scale while others find it virtually impossible to advance from their low status. The result is a widening of the global development gap.

How might economic development affect inequalities within a country?

Different regions within a single country have tended to move up and down the national development continuum. In the late medieval period when wool was England's main export, regions such as East Anglia were the most prosperous. By the mid-nineteenth century the coalfield regions were the economic 'engines' of the new industrial economy and East Anglia became a rural backwater. The regional pattern is dynamic: economic and technological changes eventually made the coalfields obsolete while the growth of services and light industries brought prosperity to rural areas. In some countries the development of tourism has favoured coastal regions rather than inland areas. An example is Mombasa in Kenya which, thanks to international tourism, has become relatively prosperous compared with most other parts of the country.

To what extent is the 'development gap' increasing or decreasing?

Key ideas	Content detail
• Some areas are finding it very difficult to develop economically. • The gap between the richer and poorer areas (those struggling to generate development) is increasing. • Some areas are able to develop rapidly and narrow the gap between them and richer areas.	The study of the concept of the 'development gap' to illustrate: • what the 'development gap' is and why it exists (including models of spatial development) • the factors (physical, economic, social, political) that may be influencing the increase or decrease of this 'gap'

Key questions

What is meant by the 'development gap'?

The development gap refers to differences in wealth between the richer countries of the developed world and the poorer countries of the developing world. The gap has increased with time. In the 1980s a gathering of world politicians led to the Brandt Report, which identified a global **North–South divide** between the affluent North and the poor South.

What is meant by the process of cumulative causation?

Gunnar Myrdal's model describes the process of circular and cumulative causation. An area possesses **initial advantages** such as natural resources or good accessibility. Economic growth creates employment, and people move in, adding to the local market as well as providing labour. **External economies of scale** mean that economic activity continues to expand, supported in part by growing tax revenues, which can lead to investment in infrastructure such as transport and education.

Due to this upward spiral of growth, a **core region** emerges at the expense of the **periphery**. The beneficial **spread effects** of growth from the core are counteracted by **backwash effects** such as the movement of commodities and labour into the core. The spatial consequence is uneven development.

What are the key elements of the Friedmann model?

John Friedmann's model describes four stages of spatial economic development:

- **Stage 1**: pre-industrial period when urban centres serve local markets but with very little interaction among them. Development potential is very limited.
- **Stage 2**: industrialisation begins as a strong core region emerges, often with a major metropolitan centre at its heart. Initial advantages result in rapid growth of the core attracting resources and labour from the periphery, which is dependent on the core. Few benefits flow to the periphery. Overall the economy is fragmented and development has a very uneven distribution.
- **Stage 3**: development spreads from the core. Although the core remains the most developed part of the system, other areas begin to develop. In part this is because other areas can offer cheaper locations for economic activity due to lower wages and land costs. In Stage 3 spatial contrasts in development are reduced as the economy becomes more integrated geographically.
- **Stage 4**: spatial inequalities are minimal. A mature hierarchy of urban places has evolved and the entire economic system is fully integrated.

What is the significance of the Rostow model?

In the early 1970s Walt Rostow proposed his **five stages of development**. The transition from a traditional, **pre-industrial** society requires some **preconditions for take-off** to exist before development can occur. This involves rising rates of investment; the growth of economic and social infrastructures such as banking and education; the emergence of an elite group of economic investors such as merchants and industrial capitalists; and a functioning centralised political system or nation state.

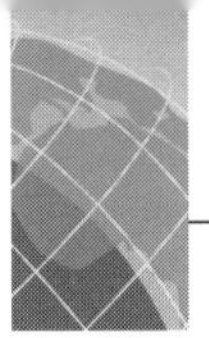

This then evolves into a period of **take-off to self-sustained growth**, a watershed period of very rapid economic growth lasting between 10 and 30 years. Self-sustained growth relies on a few **leading sectors** in the economy to drive the process forward. Once through this critical period, the **drive to maturity** stage is entered. Economic growth is now sustained and relies less on the leading sectors as the economy diversifies. Imports fall, exports rise so the balance of trade improves and high rates of investment keep the economy moving forward.

The fifth and final stage, the **age of mass consumption**, is reached when consumer goods and services are important and a strong welfare state is established.

Although not explicitly geographical, Rostow's model does have spatial implications. Leading sectors bring links to industrial regions where economic activity is localised. Some regions have the resources to nurture the development of leading sector industries. Growth in these regions surpasses growth in less favoured locations and so inequalities develop.

It is important to consider the model's origins. Rostow described his book where the first model appeared as 'a non-communist manifesto'. This helps us to understand the emphasis he placed on capitalism and the rise of consumerism. His model is based on events in the USA. But the inevitability of development based on economic advance is questionable since the development gap is becoming wider, not narrower as his model implies. In particular it ignores the important role of government and protectionism in economic development.

How do physical factors affect the development gap?

There is a degree to which a country's resource base can influence development but there are exceptions too. Countries where a large proportion of the population relies on agriculture can be adversely affected by environmental factors. Prolonged drought has had devastating impacts in the Sahel, northeast Brazil and parts of India and China. Arid and semi-arid regions have great rainfall variability, which reduces average crop yields. Many people in these regions are already among the poorest in the world and have few entitlements to draw on in times of food shortage. The problem is often aggravated by severe degradation of soil, vegetation and water resources.

Major floods are common in some regions (e.g. Bangladesh), resulting in loss of crops and livestock, damage to farmland and destruction of transport systems. It may take a country many years to recover from such a disaster, making it even more difficult to close the development gap.

How do economic factors affect the development gap?

Some parts of the developing world have benefited from globalisation, increasing per capita incomes and living standards. However, with few exceptions, globalisation has had little impact in sub-Saharan Africa. Today 20 countries in sub-Saharan Africa have lower incomes per capita in real terms than in the 1980s. Although some developing countries have made rapid economic progress, often this has been based on rising commodity prices. When prices fall, over-reliance on a narrow range of primary exports can spell economic catastrophe. Uneven and unequal development remains a feature of the global economic system. Competition for global capital is fierce and is overwhelmingly dominated by the developed world.

How do social factors affect the development gap?

Population changes can have significant influences on development. Rapid population growth in many developing countries means that these countries are having to work hard simply to keep their healthcare and education services at the level they are now, never mind increasing their availability.

A particular concern among the poorest countries is the impact of **Human Immunodeficiency Virus (HIV)** and **Acquired Immune Deficiency Syndrome (AIDS)**. Sub-Saharan Africa is most badly affected, with two-thirds of the world's HIV cases, some 22.5 million individuals. The resulting high mortality, especially among young adults, leaves millions of orphaned children in the care of elderly relatives and older siblings. The AIDS epidemic also increases **food insecurity**, causes children to miss out on schooling, and creates labour shortages that reduce production and tax revenues.

How do political factors affect the development gap?

Historically the economic progress in the developed countries has been associated with increasing emancipation and the widening of democratic involvement. In contrast, in NICs such as South Korea and China, direct government involvement has driven development.

In many developing countries poor and corrupt governance has been a major obstacle to economic progress. Some political problems, however, have their origins in colonial times. For example, many African states that were former colonies have borders that pay little regard to existing tribal loyalties. The result was the grouping of people who have little sense of nation. In the post-independence period this situation has provoked internal tension, conflict and even civil war.

In what ways do economic inequalities influence social and environmental issues?

Key ideas	Content detail
Economic inequalities may result in social and environmental conditions also becoming unequal.	The study of variations in social and environmental conditions to illustrate: • variations in pollution (such as air, water, solid, noise) in both a developed country and an NIC • the economic and social inequalities within one named region or a large city resulting from the interlinking of economic and social factors

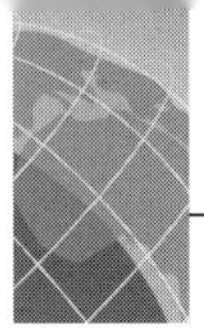

Key questions

How and why do economic inequalities result in inequalities in exposure to environmental pollution in developed countries and NICs?

According to Friedmann's model, during economic development there is often a period of time when spatial inequalities widen. As a result some groups gain more from development than others. For NICs, economic growth has priority over environmental concerns. The emphasis is on jobs and wealth creation. These tend to be concentrated in core regions and it is there that the worst environmental problems (especially pollution) are found. Environmental pollution that adversely affects health and the quality of life is a **negative externalities** or '**spill-over**' effect of industrialisation.

The metropolitan region of **Mexico City** has a high level of environmental pollution. Its population of around 21 million generates huge amounts of domestic waste, and hundreds of industries pollute the atmosphere, watercourses and soils. Vehicles are the major source of air pollution, particularly nitrogen oxide, ozone and particulates. Although environmental laws exist, they are commonly ignored. Physical geography also contributes to Mexico City's high levels of pollution. Located in a natural basin and surrounded on three sides by mountains, the city experiences frequent **temperature inversions**, which trap pollutants. Additionally, high altitude (Mexico City is 2,240 m above sea level) means that combustion is less efficient. Noxious gases and hydrocarbons released from car exhausts combine with sunlight to form a brown **photochemical smog** which hangs over Mexico City for much of the year. Air pollution is worst in slum settlements where open fires, rubbish tips and dust from unpaved roads are major sources of contaminants.

Many developed countries have a legacy of pollution from former industrial activities, such as contamination of sites by toxic waste. Despite rigorous environmental controls, many heavy industries continue to pollute and can have significant effects on local populations. Hardest hit are poorer groups, forced to live in neighbourhoods where housing is cheap due to the disbenefits of pollution. Teesside in northeast England has one of the biggest concentrations of heavy polluting industries in western Europe. Iron and steel, chemicals and oil refining are responsible for high levels of air, water, noise and light pollution. Life expectancy in areas close to these industries is significantly lower than average. Respiratory illnesses such as lung cancer and bronchitis are higher than average.

How are economic and social inequalities in a large region or city related to interlinkages between economic and social factors?

Deprivation and poverty are found in all large cities. **Socioeconomic deprivation** is related to a number of factors: low income, unemployment, poor health/disability, inadequate education and skills, poor housing, high levels of crime and so on. Taken together, these factors are combined into an index of **multiple deprivation**.

The underlying cause of multiple deprivation and the resulting inequality is economic. Low wages are common in sectors where uncertainty of employment and part-time working are the norm. People with low skill levels, often due to inadequate education, have only limited employment opportunities. Lack of money can often lead to poor education, with children leaving school early to contribute to family incomes. Parents with little formal education of their own may not see education as a priority. Poverty means that the choice of where to live is limited, and poor housing is linked to poor health (**morbidity**), absence from work and lower incomes.

With relatively few jobs in the **formal sector** in developing countries, a large part of the workforce is **under-employed** or at best has only casual employment. People often work long hours for low pay, or are self-employed in the **informal sector** (e.g. street hawking, recycling). The resulting low incomes are linked to a range of social and economic factors and multiple deprivation.

To what extent can social and economic inequalities be reduced?

Key ideas	Content detail
It is important to reduce extreme inequalities and various approaches have been tried with differing degrees of success.	The study of the management of social and economic inequalities to illustrate: • the variety of methods that can be employed to tackle social and economic inequalities and their impacts (including the use of law, education, planning, subsidies, taxation) • the reasons for, and the methods used in, reducing social and economic inequalities in one named country

Key questions

What are some of the features of inequalities?

It is generally the case that the less developed a country is, the more unequal its society. Thus some of the widest inequalities are found in sub-Saharan Africa, the world's least developed region. Meanwhile the most equitable societies are countries such as Denmark and Japan, two of the richest countries. Within all countries, inequalities tend to be associated with particular groups defined by factors such as race, ethnicity, gender and age. Inequalities arise in a number of different ways, most of which are interlinked.

What methods have been used to reduce social and economic inequalities?

At the global scale, **multilateral** organisations such as the UN and the World Bank sponsor projects designed to raise the quality of life of people in the developing world. The EU as well as individual countries like the USA also offer development aid. The **World Trade Organization** encourages free trade in the belief that removing trade barriers will benefit all.

For the past 25 years the **Fairtrade Foundation** has worked with small-scale producers in developing countries to provide them with better access to markets in the developed world. Increasingly, mainstream retailers such as the supermarket

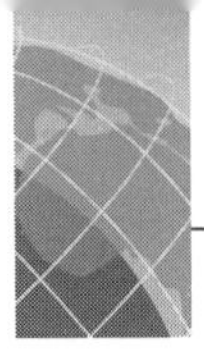

chains stock fair-trade products. Individual countries have overseas development agencies through which money, materials and expertise are channelled (e.g. UKAID and USAID).

Non-governmental organisations (NGOs) such as Oxfam, Christian Aid and Médecins sans Frontières provide important humanitarian and development aid. They are largely supported by voluntary donations from the public.

Some countries, in both the developed and developing worlds, have minimum-wage legislation in place. The provision of universal education is an important step to narrowing inequalities and lifting people out of poverty. Education has played a crucial part in the economic success of most NICs and RICs, although within countries such as China and India, inequalities still remain high. Many individual countries have government programmes aimed at reducing intra-national inequality. The taxation system is widely used to redistribute wealth. Many governments have **progressive** systems where levels of taxation increase with income. Subsidies such as free school meals and help with university fees are available to families on low incomes. Pensioners receive benefits such as cheaper travel and grants towards their winter heating costs.

Development programmes exist at both national and supra-national scales to reduce regional inequality. Grants are available to employers to relocate or expand in areas of high unemployment and retraining packages for workers are offered. Governments may also upgrade infrastructure to improve accessibility. The EU operates a **Cohesion Policy** that provides financial assistance to the poorest regions in the union. Many developed countries have anti-discrimination laws in place to reduce inequality within society on the basis of characteristics such as race, gender, age and disability.

Why have different methods to tackle inequality met with variable success?

In the past 200 years the developed world has established a dominant position in the global economy. However, the recent emergence of NICs and RICs suggests some degree of dynamism in the system. Even so, some parts of the global system such as sub-Saharan Africa seem irretrievably poor. And nationally, most countries have regions where a lower quality of life is very difficult to improve. In fact spatial inequalities are remarkably persistent at all scales. A complex array of factors — economic, social, political, economic and environmental — that operate both independently and together, explain the variable success of attempts to reduce inequalities.

Questions
& Answers

Section A: data-response questions

Section A contains students' answers to six data-response questions. There is a question for each of the six optional topics in Unit F763. These questions are based around data presented as tables, charts, maps, newspaper clippings etc. All the questions have exactly the same wording:

'Outline a geographical issue indicated and suggest appropriate strategies for its management.'

While the stimulus materials change for each examination, questions will retain the same wording. In Unit 763, each data-response question is worth 10 raw marks, and overall this part of the assessment accounts for one-third of the available marks for the unit. In the examination, you must answer three data-response questions, with at least one chosen from **Environmental issues**, and one from **Economic issues**. You should allow yourself no more than 20 minutes for each question and aim to complete Section A in around 60 minutes.

Section B: extended writing questions

Section B has students' answers to six extended writing questions. In the examination you must answer two extended writing questions, choosing one question from **Environmental issues** and one from **Economic issues**. The extended writing questions deal with environmental and economic issues and are therefore discursive. They are also **synoptic**, allowing you to draw on your wider geographical knowledge and understanding and to make connections with other areas of the specification you have studied. Each extended writing question is worth 30 raw marks. This suggests that you should allow about 45 minutes per question.

Examiner's comments

Examiner's comments, indicated by the *e* icon, follow each student answer. These comments show how the marks have been awarded and highlight areas of credit and weakness. For weaker answers the comments suggest areas for improvement, focusing on specific problems and common errors such as lack of development, excessive generalisation and irrelevance.

Section A: data-response questions

Environmental issues

Earth hazards

Resource 1

The 7.6 magnitude Bhuj earthquake that shook the Indian Province of Gujarat on the morning of 26 January 2001 was one of the two most deadly earthquakes to strike India in recorded history. One month after the earthquake official government of India figures placed the death toll at 19,727 and the number of injured at 166,000. Indications were that 600,000 people were left homeless, with 348,000 houses destroyed and an additional 844,000 damaged. The Indian State Department estimated that the earthquake affected, directly or indirectly, 15.9 million people out of a total population of 37.8 million. More than 20,000 cattle were reported killed. Government estimates placed direct economic losses at US$1.3 billion. Other estimates indicate losses may be as high as U$5 billion.

Resource 1 focuses on the Bhuj earthquake in Gujarat in India, in January 2001. Outline a geographical issue indicated and suggest appropriate strategies for its management. (10 marks)

B-grade answer

One possible issue is why the Bhuj earthquake was so devastating and could its effects have been mitigated. The quake killed nearly 20,000 people, injured 166,000 and left 600,000 people homeless. Even though it was powerful (7.6M), such a level of destruction and so many deaths and injuries may seem excessive. In other words did the level of destruction owe more to human than to physical factors?

Unlike some natural hazards such as hurricanes and volcanic eruptions, earthquakes cannot be predicted. However, there are a number of management strategies that can be adopted to reduce the death and destruction caused by earthquakes. Hazard management in Gujarat could have reduced the vulnerability of the region to earthquake disasters. People could be educated in how to respond to a major earthquake and more training could be given to the emergency services to coordinate an effective response to a disaster. This would include the emergency provision of

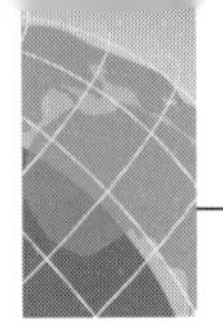

water, food and medical supplies. Both of these responses would help to reduce the death toll if a similar earthquake hit the Bhuj region again.

Because the main loss of life in earthquake hazards is caused by the collapse of buildings, seismic building codes may be developed. These would involve building regulations (e.g. building materials, reinforcement, fire proofing) to prevent collapse and fire. It is particularly important not just to have building codes (most countries exposed to earthquakes hazards have them) but for governments to ensure that they are applied in practice. Making a society more resistant to earthquake hazards by constructing earthquake-proof buildings involves long-term planning and huge investment.

The answer identifies an issue that often arises in the aftermath of major earthquake disasters. There is good knowledge and understanding of a range of possible management strategies though their evaluation is limited. Good use is made of geographical terms, although the evidence that ideas and concepts have been drawn from a wider area than hazard geography is not convincing. Although quite detailed, the answer does not fulfil all the criteria for a Level 3 answer. It is therefore assessed as a B grade (7/10 marks).

Ecosystems and environments under threat

Resource 2 Plant and animal species at risk in the Nevada Desert spring communities

Native spring-dwelling species in southern Nevada	Endangered species/ sub-species	Threatened species/ sub-species
Mammals	1	0
Birds	2	0
Fish	11	2
Plants	1	2

Resource 3

Explosive economic and population growth in Las Vegas, Nevada, has stimulated demand for additional water supplies. However, water demand in this arid environment has reached the limits of current supply. To meet future needs, local officials from Las Vegas and satellite communities hope to obtain rights to about 1.32 billion m^3 per year from a regional groundwater aquifer extending from Salt Lake City, Utah, to Death Valley, California. This aquifer feeds the Great Basin spring systems. These springs support unique ecosystems of great biodiversity, including 20 species and sub-species — mainly snails, insects and fish — listed under the Endangered Species Act.

Source: *Bioscience*, September 2007, Vol. 57, No. 8.

Resources 2 and 3 relate to a proposal to extract groundwater from a desert aquifer in southern Nevada. Outline a geographical issue indicated and suggest appropriate strategies for its management. (10 marks)

A-grade answer

There are a number of issues that might arise through the extraction of groundwater from a desert aquifer. However, the main one relates to the damaging impact of the development on wildlife in habitats around desert springs and whether in the long term water could be extracted sustainably.

The springs in the Great Basin fed by the aquifer support unique ecosystems of great biodiversity. If this water source declines, up to 20 species listed as endangered may become extinct. This in turn could have knock-on effects on food chains and nutrient cycles and badly affect other species in the ecosystem.

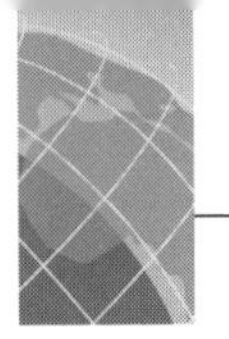

One possible management response would be to reduce the demand for water from big cities like Las Vegas. This could be done by increasing water charges for domestic and business users. While unpopular, higher charges would almost certainly reduce demand. Management could also acknowledge that there must be limits set by water resources to population and economic growth in a desert region. Action could be taken to limit growth (e.g. higher taxes on businesses and property) though in a free market economy like the USA, such controls might prove difficult to enforce. The planning authorities could also promote water conservation. They could limit non-essential water use such as irrigation of golf courses, parks and gardens or place high tariffs on non-essential uses like swimming pools. In some desert cities in the USA households have been given financial incentives to replace garden lawns with desert gardens planted with native species such as cacti and yucca. Managers could also look to import water from neighbouring areas which have wetter climates and water surpluses. This, however, would be costly because it would involve the construction of pipelines or canals.

e The answer suggests a realistic ecological issue that could arise through the extraction of groundwater. An attempt is made to develop the nature of the issue by referring to some of the stimulus material in the question. A coherent and logical set of possible management responses are outlined. They are supported with geographical information derived from other parts of the A-level geography specification (e.g. urban growth, managing supply and demand for water) and there is, in addition, some clear (if limited) evaluation of the various options. Overall the answer is well written and uses geographical terminology accurately. The answer satisfies most of the criteria needed to achieve Level 3, hence the award of an A grade (8/10 marks).

Climatic hazards

Resource 4 Track followed by Hurricane Ike across the Caribbean and Gulf of Mexico

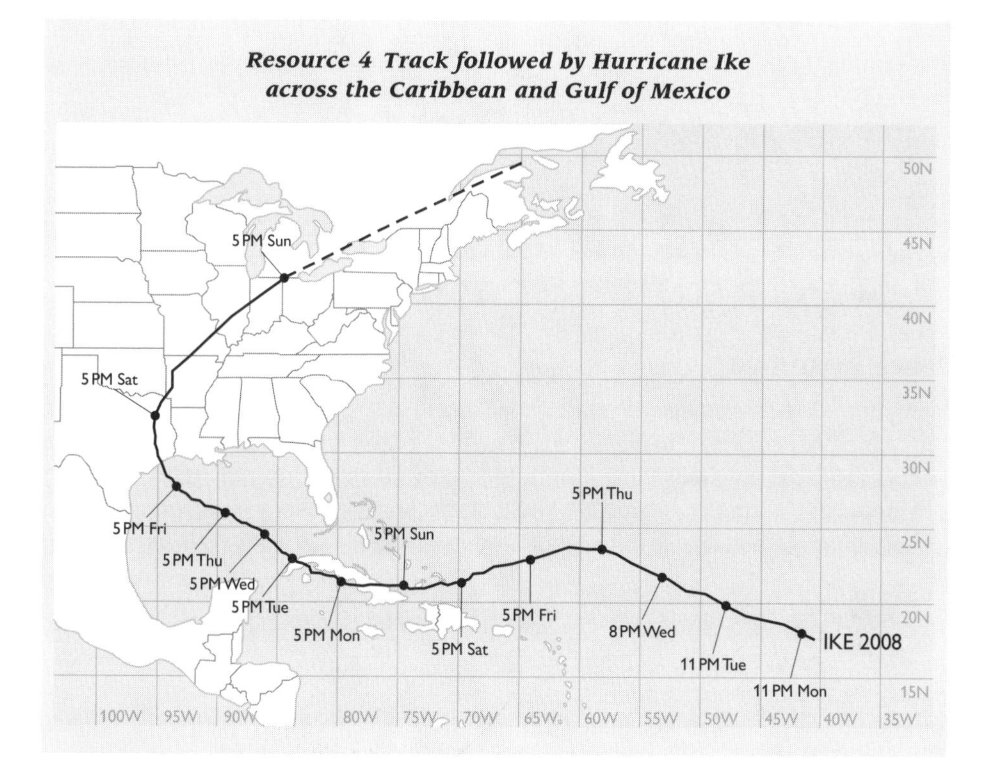

Resource 5

The Cuban government ordered 1.2 million people to seek safety with friends and relatives or in government shelters. In Havana, where Hurricane Ike was expected to unleash heavy winds and rain this morning, evacuations began in earnest yesterday afternoon. The government closed schools and government offices in the capital as people reinforced windows, removed plants from balconies and formed long queues at bakeries.

Gustav tore across western Cuba as a Category 4 hurricane last month, damaging 100,000 homes and causing billions of dollars of damage. But no deaths were reported as a result of mandatory evacuations of at least 250,000 people. 'In all of Cuba's history, we have never had two hurricanes this close together,' said José Rubiera, head of Cuba's meteorological service. Waves created by Ike crashed into apartment buildings, hurling heavy spray over their rooftops, and winds uprooted trees. Falling utility poles crushed cars parked along narrow streets in the central city of Camagüey and the wind transformed buildings of stone and brick into piles of rubble.

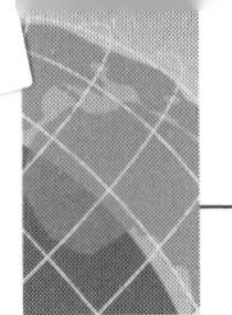

Felix García, a meteorologist at the US National Hurricane Center in Miami, forecast that Ike would become even more powerful. 'It's over warm waters,' García said. 'It can definitely maintain its strength right now, and when it's out of Cuba it has the potential to become a lot stronger.'

Source: newspaper extract from 9 September 2008.

Resources 4 and 5 relate to Hurricane Ike, an Atlantic hurricane that hit the Caribbean and Texas in September 2008. Outline a geographical issue indicated and suggest appropriate strategies for its management. (10 marks)

A-grade answer

A possible issue connected with Hurricane Ike and other hurricanes in the Caribbean and Gulf of Mexico is how to tackle the problem of coastal flooding from storm surges in the most exposed areas close to sea level. For example, most of the destruction and loss of life in this region in 2005 (the worst hurricane season on record) was due to storm surges and subsequent coastal flooding.

An appropriate way to manage deadly storm surges would be to protect major cities (like New Orleans) and towns. This could be done by building flood embankments or levées. However, in 2005 the levées protecting New Orleans were breached by Hurricane Katrina's storm surge and so new levées could only provide protection from the worst storms if they were built much higher. This would be very costly and may not even give complete protection from Category 5 storms.

As an alternative the storm surge hazards can be managed using an ecological approach rather than hard engineering. Instead of keeping rivers in their channels (by building levées) and preventing river floods, rivers could be allowed to flood. In this way natural deposition would raise the height of the coast and reduce the impact of storm surges. In Bangladesh, planting mangrove forests along the coast would have a similar effect and also help to reduce the power of storm surges. These ecological responses are relatively low-cost solutions and they work with natural processes.

Where evacuation is impossible (e.g. many parts of coastal Bangladesh and Myanmar) the construction of storm shelters can save thousands of lives. They act as temporary refuges and are elevated above the floodwaters on stilts. They have been highly successful in Bangladesh since following major storm surge disasters in 1971 and 1991.

e The answer selects a valid geographical issue connected with hurricane disasters. It briefly explains the context of the issue and then proceeds to consider three or four possible management responses. The suggestions demonstrate a good knowledge and understanding of the storm surge hazards and management strategies. There is sound evaluation of these strategies reinforced by appropriate exemplification. Although there is some connection with aspects of ecology drawn from elsewhere in the specification, a general lack of explicit synopticity is a weakness. Apart from this caveat, this is a Level 3 answer which therefore achieves an A grade (8/10 marks).

Economic issues

Population and resources

Resource 6 Water resources (m³ per person per year)

	1988–92	1993–97	1998–2002	2003–07
China	2,364	2,251	2,165	2,125
Egypt	31.1	28.3	25.7	24.3
India	1,426	1,299	1,196	1,142
Mexico	4,668	4,277	3,973	3,821
Pakistan	468	415	370	348
Uzbekistan	760	690	642	614

Resource 6 relates to changes in water resource availability in selected countries between 1988 and 2007. Outline a geographical issue indicated and suggest appropriate strategies for its management. (10 marks)

D-grade answer

There is a decline in water resources in all the countries on the table. China has gone down from 2,364 to 2,125 and India has gone down from 1,426 to 1,142. All the other countries have also gone down in their water resources. This means that they will not have enough water for their uses in the home and for industry. They will probably have to use polluted water instead which can mean all sorts of issues. Polluted water spreads diseases such as cholera and causes many deaths especially among babies. The way to solve this is to have better water supplies which can be cleaned. One way to do this is to build reservoirs to store water. The Chinese have built a massive dam the 3 Gorges Dam which has created a reservoir hundreds of miles long. The water can be used to supply millions of Chinese with clean water as well as electricity. This means the Chinese will have more water which will allow them to raise their standard of living.

Mexico is more developed than China and can afford to have more water supply. However, it is still loosing water as in 1988–92 it had 4,668 and in 2003–07 it had 3,821. This may be because they have had a draught and so they need to store water when it rains. One way to do this is to build dams which can then supply water in a draught. This way water can be supplied even when there is no rain. The problem with building dams is that you have to take the land from people and they have to be moved. The Chinese have moved millions of people to new towns when the water built up behind the 3 Gorges dam. Another problem with dams is that they can cause problems for the ecosystem. There is a rare dolphin that lives in the river below the dam and this has been made extinct because the dam has changed the water flow of the river.

There are aspects of this response that are relevant but others that are not. It begins with a suitable summary sentence and then quotes some figures, although the units should be given. An appropriate geographical issue is identified — lack of water for domestic and industrial use. The link with polluted water is fine but is not explained and is unconvincing. The management strategy of dam building to form reservoirs is appropriate and the example quoted relevant. More details regarding its use as a supply of water for consumption would make it a more valuable part of the answer. The material on Mexico is not really secure as it lacks any real world information. There are also some spelling mistakes. Overall this is an answer with some merit but not one that is more than the bottom of Level 2. It would be awarded a D grade (5/10 marks).

Globalisation

Resource 7

In southern India's Tirupur town, young girls are lured to work in the garment industry with a promise of a 'golden opportunity' to earn their own dowry at the end of a 3-year apprentice period. Known as 'camp coolies', they instead end up working in deplorable conditions for years getting virtually nothing.

While NGOs say labour laws are not stringent enough, Tirupur Exporters say that flexible labour laws favourable to entrepreneurs are needed if India is to compete with China, whose market share of the garment exports industry is 25–27% as opposed to India's 3.5%.

According to a Tirupur People's Forum (TPF) study, these girls are paid between 70 and 95 cents a day. Every month, US$9.5 to US$11.5 are deducted for boarding and lodging. According to the NGO SAVE, these girls are almost like prisoners in their hostels, which are usually in the same compounds as their workplace, and can only step outside the gates escorted by a warden.

Highlighting the deplorable housing conditions of these workers, the TPF study says: 'Sumangali scheme workers are kept in an abandoned poultry farm. Approximately 50 to 60 women workers sleep in a 25 × 6 square metre area, which is dusty and dingy. There is no space for privacy and there are only four toilets.' Many suffer severe nutritional deficiencies as well.

Resource 7 provides information on some aspects of globalisation. Outline a geographical issue indicated and suggest appropriate strategies for its management. (10 marks)

A*-grade answer

Globalisation has involved many manufacturing industries moving from their base in developed countries such as the USA and Germany to developing countries and NICs. Countries like India but also Taiwan and Malaysia have had many industries set up there.

There are advantages and disadvantages of globalisation as this article suggests. There are the wages that local people receive, however, very often these wages are very low. In this case the girls only get between 70 and 95 cents a day. The low labour costs are what attract manufacturing firms as these are much higher in developed countries. There are no additional costs such as pensions or sick pay and in this case, money is deducted for boarding and lodging.

One strategy is to have minimum wages in places like India. The government can pass laws to make employers pay a decent wage. This is not easy to enforce though as the local factory owners can bribe the local police to ignore what they are doing. It is difficult as well because many governments in developing countries do not have enough staff to inspect the factories. One way of helping solve this problem is if the TNC trading with the local factory owner do their own inspection and insist on the wages being reasonable.

Another possible strategy is for consumers in developed countries who buy these goods to ask for Fair Trade goods. Fair Trade means that the producers in developing countries receive a fair price for their products. Mostly this is done with foods such as tea and bananas but it is also done with some clothes. Ethical trading is growing in developed countries and should help stop exploitation of workers.

Another strategy is for goods to be traded fairly around the world. The WTO is trying to get agreements about international trade so that there are not subsidies in some countries which mean that their goods are cheaper than other countries who don't have subsidies. Also they want to stop tariffs being used to make exports more expensive which protect the producers in the country using tariffs.

This is a focused and thoughtful response. After a sensible and brief introduction, the student has kept the issue of low wages in mind throughout the answer. The resource has been used directly in a couple of places but the answer considers the issue in broader terms. The various strategies are appropriate and outlined in sensible detail. This is a Level 3 response deserving full marks and therefore an A* grade (10/10 marks).

Development and inequalities

Resource 8 Changing inequalities between countries

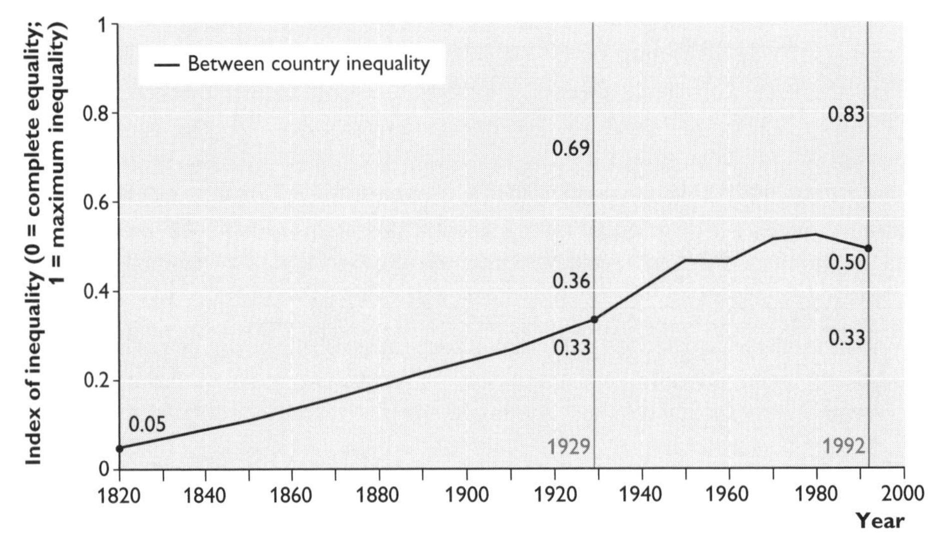

Source: *UN World Development Report*, 2006

Resource 8 provides information on inequalities between countries. Outline a geographical issue indicated and suggest appropriate strategies for its management. (10 marks)

B-grade answer

There are big differences in the world today in economic and social terms. The developed countries have nearly all the wealth for a small proportion of the world's population. Places like the USA and Japan are very wealthy with high GDP. At the other end of scale are the developing countries like in Africa, where GDP is much lower. Countries like Kenya and Gambia are much poorer. They don't have as much money to spend on healthcare and education and roads so life for their populations is very hard. There is a big difference between these countries. Some countries have improved their economy such as the NICs. These are countries like South Korea and Taiwan. They have had rapid economic growth with many industries developing for example electronics. GDP in South Korea is about $20 000. Their living standards have improved so that they are closer to developed countries than the developing countries. So the development gap between the rich and poor has grown. There are many development projects to try to make things better in developing countries. International aid is given by developed countries to developing countries for projects such as providing clean water and health clinics. Both government aid and money from NGOs like Save the Children are used to improve living conditions. The World Health Organization has had a campaign to eliminate smallpox and this has basically been done. They paid for millions of people to be vaccinated. Because fewer babies were dying the fertility rates have started to go down and this will help development. With a smaller family the children are likely to be better fed and so they are less likely to be ill. This means they miss less school and will be better educated. They are more likely to get a better job and so will have a higher income.

e This is an interesting response which makes some valuable points but taken as a whole does not totally convince. Nowhere does it refer directly to the resource although the student has correctly identified the key issue of inequalities between countries. It establishes these differences in terms of the groupings developed countries, developing countries and NICs, with suitable examples — which is fine. It makes the valid point about improvements in NICs compared to other countries. The comments about aid are relevant and details on the effect immunisation can have are helpful. There are, however, no paragraphs and in places communication could be clearer. This is a Level 2 response and would be awarded a B grade (7/10 marks).

Section B: extended writing questions

Environmental issues

Earth hazards

To what extent can preparedness and disaster planning mitigate the effects of earth hazards?

A-grade answer

The preparedness of a country or society to withstand earth hazards such as earthquakes refers to such things as the resistance of buildings to collapse and the effectiveness of emergency relief operations. These and other factors such as poverty influence the vulnerability of a country or society to earth hazards. Japan for example is a rich country and so can invest in making society earthquake-proof and implement disaster planning. China, by comparison, does not have the resources to protect most of its population against hazards such as earthquakes and this may explain why 87,000 people were killed in the Szechuan quake in 2008.

For a country to be well prepared for an earthquake hazard the population must be educated in emergency drills and contingency plans for evacuation, and the provision of emergency relief must be available. Also in earthquakes, buildings must be strengthened to prevent collapse, and to combat river floods structural engineering to control floodwaters (e.g. levées, dams) are needed. Preparedness can also be increased if attempts to predict hazards such as volcanic eruptions and floods are accurate. However, earth hazards such as earthquakes continue to defy accurate prediction.

In January 1994 a 6.7M earthquake, whose epicentre was in Northridge, struck Los Angeles. It was caused by movement along a transverse thrust fault linked to the 'big bend' in the San Andreas fault that separates the Pacific and North American plates. The quake happened in densely populated southern California, in a region of high earthquake preparedness. Buildings were reinforced so that even close to the epicentre few buildings collapsed. However, not all buildings remained undamaged. Freeway intersections collapsed at several sites and 170 bridges sustained varying degrees of damage. Some steel-framed buildings, specifically designed to be seismic resistant, cracked and reinforced concrete columns were crushed. But on the whole the region was well prepared and the response to the quake was immediate and successful. The death toll was just 57. Even so, the Northridge quake was the costliest in US history, with damage to homes, public buildings, freeways and bridges exceeding US$20 billion.

The Kashmir earthquake, which occurred in northern Pakistan and northern India in October 2005, hit a region where the level of preparedness was low. Its epicentre was 100 km north of Islamabad, near the town of Muzzaffarabad. The quake was caused by the Indo-Australian plate subducting under the Eurasian plate and causing powerful earth movements along low-angled thrust faults. The quake measured 7.6M and its focus was shallow — just 26 km deep. Exposure was high not just because of the nature of the quake but also because it occurred in a region where 15 million people lived. These people were vulnerable to earthquakes because of widespread poverty (one-third of the population lives below the poverty line, most of them in rural areas). Moreover, governments have spent little on earthquake mitigation. Most buildings were not earthquake-proof. Those in rural areas typically were made of mud and cobbles, held together with weak mortar. Violent shaking simply led to their collapse, trapping and killing their inhabitants.

Neither the Pakistani nor the Indian government had planned for emergency relief and the response to the disaster was very slow. Landslides blocked many roads, making it impossible to reach victims without helicopters. There was a shortage of tents and clean drinking water. Overall, half of all buildings in Muzzaffarabad were destroyed and there was extensive damage to roads and other infrastructure. The cost of repair was estimated to be US$5 billion. Three million people were made homeless and 87,000 people died.

Although Kashmir suffered far more deaths than Northridge, the cost of damage to buildings and infrastructure was far lower. In developed countries there are far more expensive buildings to repair and reconstruct and insurance pay-outs are much higher than in developing countries. The comparison shows, however, that a greater preparedness in an earthquake-prone region can prevent major loss of life. Even so, despite all the precautions taken in southern California, many buildings failed. It is also possible that an excessive reliance on structural solutions to the earthquake hazard may encourage people to continue to live in hazardous zones when other locations would be preferable. Although there will always be some damage, providing building regulations and disaster planning are implemented, it seems that a 'prepared' society will be far better equipped to withstand a major earthquake.

Volcanic eruptions are another example of an earth hazard. Differences in preparedness are evident in the eruptions of Mount St Helens in the USA, and Mount Pinatubo in the Philippines. In June 1991 Mount Pinatubo erupted and sent an ash cloud 35 km into the atmosphere. Around 300 people lost their lives, mainly due to lahars and the collapse of heavy roofs under the weight of wet ash. The official response to the eruption was rapid, with 60,000 people being evacuated from the slopes immediately before the eruption. However, post-disaster planning was less effective and around 500 people died as a result of poor conditions in the refugee camps. The eruption of Mount St Helens in May 1980 provides a contrast. Preparedness was much greater,

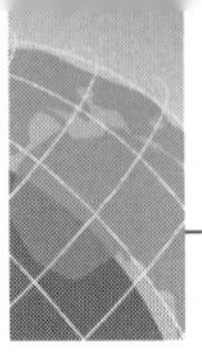

with the volcano being continuously monitored for signs of eruption (e.g. earth tremors, ground inflation, gas emissions). As a result early warnings were broadcast and most people were evacuated from the danger zone. There were, however, 57 deaths, though this was less than one-fifth of those killed in the Pinatubo disaster.

Structural methods of flood control are often used to prevent floods but can sometimes create problems of their own. Levées allow rivers to discharge more floodwater, but if they are breached the damage they cause can be much greater than a normal flood. The floods on the Mississippi River in 1993 were made worse by the levée system. Channel straightening can also cause problems. The river responds by scouring its channel, with eroded material then deposited downstream. Aggradation in this area may shallow the river and increase the risk of flooding.

Thus preparedness does not necessarily solve hazardous events like flooding, and in some instances can even make matters worse.

This is a thorough account which demonstrates good levels of knowledge and understanding (AO1). It is written around detailed examples that are appropriate and, as a result, avoids excessive generalisation. A range of earth hazards is considered, though there could be better balance, with less emphasis on earthquakes. Cause and effect are well understood throughout. For knowledge and understanding the answer would achieve Level 3 and 8/9 marks.

Analysis, evaluation and interpretation are slightly less convincing. Some evaluation occurs throughout the answer but it could be more explicit and more obvious in places. References to tectonic processes, population distribution and the economic status of societies suggest linkage with other parts of the specification, and an attempt to write broadly and synoptically. The answer would reach a good Level 2 for AO2 and scores 13/17 marks.

For investigation, conclusion and communication (AO3) the answer also reaches Level 2. The account is well structured, with accurate spelling and grammar, and good use of geographical terminology. The answer fails to achieve Level 3 because a clear conclusion at the end of the account is omitted. There are, however, attempts to provide some summary within the main body of the answer. For AO3 the answer would score 3/4 marks. The final overall total for this answer would be 24/30, which would just gain an A grade.

Ecosystems and environments under threat

Discuss the view that humankind's current exploitation of natural ecosystems is unsustainable.

A*-grade answer

Sustainability means meeting the needs of the present without compromising the needs of future generations. Natural ecosystems offer valuable resources to society but they are often fragile and their survival is delicately balanced. Humankind's current exploitation of these systems may be thought to be unsustainable. Thus, broadly speaking, the statement is probably true. This can be shown by reference to a number of examples at different scales.

The exploitation of fossil fuels such as coal and gas is unsustainable if only because these resources cannot be replaced except over millions of years. Burning fossil fuels threatens the global ecosystem. It is the main cause of global warming and climate change. Already global warming is having major ecological effects in the Arctic and sub-Arctic — melting permafrost, allowing the coniferous forests to extend into the tundra and destroying habitats. Many animals may face extinction this century. At the same time the Arctic Ocean is becoming more acidic as ocean waters absorb carbon dioxide, damaging animals at the base of the food chain and ultimately threatening fish and marine mammals. At a similar large scale, the tropical rainforests are being destroyed and degraded by commercial timber operations, mining and large-scale ranching. Destruction of the forests releases carbon dioxide, which enhances the greenhouse effect and increases global warming. In no sense is this exploitation sustainable.

Unsustainable exploitation also occurs at more local scales. Overexploitation of limited ecological resources (e.g. timber, grazing, water) in countries such as Mali, Niger and Burkina Faso in west Africa has caused widespread land degradation. In the Nara region of Niger rapid population growth has led to overstocking with excessive numbers of cattle and goats exceeding the land's carrying capacity and causing deforestation, overgrazing and soil erosion. Together with drought, the loss of resources through the unsustainable exploitation of the local ecosystem has led to widespread famine, food shortages and out-migration.

Thorne and Hatfield Moors in South Yorkshire provide another local example of unsustainable exploitation of natural ecosystems. These lowland peat bogs support a unique wetland ecosystem and some of Britain's rarest plants and animals (e.g. spiders, dragonflies, grass snakes, nightjars). However, for many years the peat has been mined on an industrial scale, mainly to provide organic material and compost for gardening. Thousands of tonnes have been dug up, deep trenches have lowered the water table, and a large part of the moorland ecosystem has been completely destroyed. Today only 6% of the original wetland remains. However, what remains of the moors is now being conserved for wildlife and recreation, suggesting that human exploitation of natural ecosystems is not always unsustainable. Some land which is

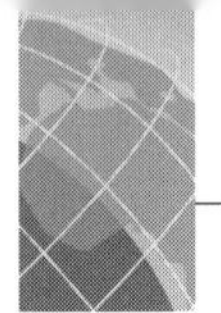

not too badly damaged is also being rehabilitated and water tables are being restored to their original level. Current use of the moors is for the first time in many years sustainable.

There are many other examples of the sustainable exploitation of natural ecosystems. Jasper National Park in the southern Rockies in Canada, famous for its spectacular scenery and wildlife that includes grizzly bears and wolves, has UN World Heritage Site status. The priority of the park is to protect the natural ecosystems, and to allow human use of the area (i.e. mainly through tourism) that is sustainable and does not impair wildlife and habitats. Examples of management approaches include restricting parking and imposing traffic quotas in the most sensitive areas of the park like Mount Edith Cavell.

However, in conclusion I believe that the statement is largely correct and that unplanned exploitation of ecosystems inevitably results in damage that may be irreversible. The general weight of evidence suggests that at all scales the current exploitation of natural ecosystems is unsustainable. But on a more positive note, progress in the sustainable use of natural ecosystems is beginning to take place as society's awareness of their value and importance increases.

This is a balanced answer which, through its use of a range of detailed examples, shows good levels of knowledge and understanding (AO1). Examples are selected at different scales and in a number of different geographical locations. Linkages between cause and effect relationships are clearly understood. For knowledge and understanding the answer would achieve Level 3 and 8/9 marks.

Analysis, evaluation and interpretation (AO2) are clear, and the answer adopts a mainly discursive approach, recognising that the unsustainable exploitation of natural ecosystems is by no means universal. References to tourism, National Parks, land degradation and the effects of climate change demonstrate an attempt to widen the discussion and make connections with topics studied elsewhere in the specification. Overall the answer would reach Level 3 for AO2 and score 15/17 marks.

For investigation, conclusion and communication (AO3) the answer would also reach Level 3. The answer is well structured, with effective spelling and grammar and geographical terminology is used accurately. In addition, there is a clear conclusion, which is consistent with the argument presented in the main body of the answer. For AO3 the answer would reach Level 3 and score 4/4. Overall this answer would gain 27/30 marks and suggest an A*grade.

Climatic hazards

'Global climate change poses the greatest risks to the world's poorest countries.' Discuss.

B-grade answer

Climate change poses many risks both to rich and poor countries. Among the risks are rising sea levels, increases in drought, more frequent storms and the spread of pests and diseases. Compared to rich countries, poor countries lack the economic resources to protect themselves against climate change. People in poor countries often have few entitlements which can help them overcome natural disasters and environmental change. Poverty is linked with overdependence on primary activities like subsistence farming and fishing which are especially susceptible to environmental change.

16% of the world's population could be hit by water shortages caused by climate change. Large parts of North America, South America, southern Europe and Africa will experience lower rainfall and more frequent droughts. In mid-latitudes, higher temperatures will speed up the atmospheric circulation and severe storms will become more frequent. They will increase coastal erosion and coastal flooding. In the tropics and sub-tropics warmer ocean waters will generate more powerful hurricanes and increase the number of category 5 storms. Higher temperatures and higher humidity are likely to increase the spread of tropical diseases such as malaria. The mosquitoes that spread the disease can only survive in warmer conditions. With higher temperatures forecast in future, the spread of the disease into southern Europe is possible.

Although these problems caused by climate change affect all countries regardless of their economic status, the situation is most serious in the poorest countries which lack the resources to tackle them successfully. Niger in west Africa is one of the poorest countries in the world. According to the UN, in 2007 there were only two other countries that were poorer. Since the 1960s rainfall in Niger and other countries in Africa's Sahel has declined by 30%. Drought, possibly caused by global warming and climate change has become an increasing problem. Because farmers are too poor to invest in irrigation or drill bore holes for water, crops fail and pastures become overgrazed. Large areas of land have become degraded and desertified. Some have been abandoned altogether. Elsewhere food production is not enough to support farming communities. In recent years food shortages, famine, malnutrition and even starvation have become familiar problems.

Bangladesh is another example of a poor country that has been badly affected by climate change. It is one of Asia's poorest countries. In 2005 its GDP per capita was only US$423. Tropical cyclones have become more common in the Bay of Bengal, with storm surges up to 10 m high. Forced to live in hazardous coastal areas at sea level and at extreme densities, hundreds of thousands of people have died since the early 1970s, and vast areas of farmland polluted by salt water have been abandoned. Although it's not entirely clear that these disasters are directly related

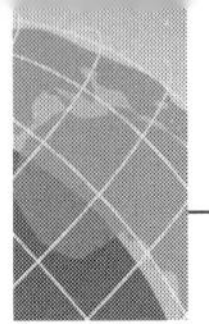

to global warming, the current sea level rise, which threatens to flood large areas of Bangladesh, almost certainly is. If this happens the people will lose everything and they will become environmental refugees. It seems unlikely that Bangladesh, given its impoverished state, will have the resources to protect its people by building levées to keep the sea out.

It is also true that global warming and climate change affect rich countries. However, the risks in these countries are less. For example, southern Spain and the US Midwest may become too dry for arable farming. However, relatively few people work in farming in these countries, and food crops could be imported from elsewhere. It may also be possible to tap groundwater supplies to irrigate crops. Loss of farmlands in rich countries is not a matter of life or death as it often is in the poorest regions of Africa and Asia.

To conclude, global climate change creates risks in both rich and poor countries. However, due to a lack of resources, populations in poor countries are much more vulnerable and governments will be able to do little to mitigate the effects of increases in droughts, storms and the spread of pests and diseases. Thus poor people and poor countries are likely to be hardest hit by global climate change.

This is a straightforward and largely successful answer to the question. There is secure knowledge and understanding of the risks posed by climate change (though with a slight tendency to overstate the role of climate change in climate hazards), supported by a number of solid examples. Cause and effect are well understood. For knowledge and understanding this answer would be marked at the top end of Level 2 and score 7/9 marks.

There is some analysis, evaluation and interpretation (AO2), and the answer provides effective discussion in a number of places, showing that both rich and poor countries face problems of climate change. The point is made that differences tend to be ones of variable severity of impact. Attempts are made to engage with other relevant topics, such as subsistence economies, agriculture and desertification, which feature elsewhere in the specification. However, discussion could be more prominent, possibly at the expense of lengthy descriptions of climate change. Again the answer would gain an upper Level 2 for AO2 and score 12/17 marks.

For investigation, conclusion and communication (AO3) the answer would reach Level 3. The answer is structured, and spelling and grammar are used effectively. There is also confident use of geographical terminology. The answer ends with a brief conclusion that restates the main point of discussion, and follows logically from the argument presented in the answer. For AO3 this answer would gain Level 3 and score a maximum 4/4. Overall, the total for this answer would be 23/30, gaining it a very good B grade.

Economic issues

Population and resources

Assess the extent to which resources are defined by technology.

C-grade answer

Resources are classified into renewable and non-renewable. Renewable resources can be used again and again as they cannot be used up. Wind power is a good example of a renewable resource as it comes from the wind and is not affected by man. Whenever it is used there is always more available. Solar power is another type of renewable resource as the sun will keep burning for millions of years to come. Non-renewable resources are used up and not replaced. When coal is burned to make power, it cannot be replaced as it takes millions of years to make coal. Coal is made from fossilised plants that have been pressurised together underground.

There are many types of different resources but some have not always been used as resources. Some resources were unknown for thousands of years as they hadn't been discovered. An example of this is uranium as it was not discovered until the twentieth century. Uranium is a mineral ore that was used in making pottery glazes. When radioactivity was discovered uranium became useful. Nuclear power was a new technology that made uranium a resource that was now useful. Uranium is mined in Canada and Australia and causes a lot of pollution. If there is a leak at a nuclear power station then radioactivity leaks into the air and this is very dangerous. In Russia a power station had a fire which caused an explosion which meant lots of people died and were affected by radioactivity. This caused cancers and birth defects and the area had to be evacuated.

Another resource like uranium is oil. It was not discovered for hundreds of years although in some places the local tribes were using it to burn but they did not know what it was. Then technology changed and oil became useful so today it is one of the world's biggest industries using lots of technology. Technology is used to get the oil from the ground and the way it is transported. Supertankers are a new type of technology that means vast amounts of oil can be transported around the world. This means that oil can be used more as a resource. Technology is also used to make oil useful as it is made into many different products such as petrol and paint. Some oil reserves have been discovered only recently as technology has advanced meaning that they could now be used. The North Sea is a hostile area to drill for oil and the technology had to advance before it could be explored. There is oil in much deeper water but this cannot be used at present as the technology is not available to get at it, for example off the coast of western Scotland.

Water power has been used for centuries but it could not produce large amounts of power until large dams could be built. Technology makes this possible, for example

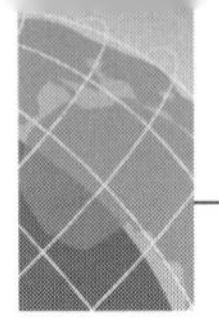

the largest dam in the world in the Three Gorges Dam in China which generates vast amounts of electricity. The river has always been there but it was not used for power until the technology was developed.

New resources are always being discovered but often it is because the technology has advanced which allows the resource to be exploited. A resource is only useful if it can be used which means it has to have the technology to extract it.

The answer shows some knowledge and understanding of the topic and although the focus stays with resources and technology, it never really discusses the question. The opening paragraph sets the scene in an appropriate way — the basic distinction between renewable and non-renewable resources. However, it does not give any indication that the student has read the question carefully and realised where the discussion should be heading. The second paragraph does bring the response back towards the question with its discussion about uranium and hints at the role nuclear power technology has had. But the paragraph soon becomes distracted with comments about the hazards nuclear power can pose. The section on oil again moves towards the question but not securely. It is good that technology is considered in terms of the transport of oil as well as obtaining it, but as with nearly all the points in the answer, too little detail is given as to how technology has influenced resource definition. The paragraph on water power has potential but it is not developed. There is an attempt at a conclusion but this does not convince. The answer would achieve 7/9 for knowledge and understanding (AO1) and 3/4 for communication (AO3). The main weakness is the absence of any real evaluation. Thus for (AO2) it would only reach Level 2 and score 9/17. Overall, this answer would gain 19/30 marks, which would be a grade C.

Globalisation

Discuss the view that the impact of inward investment by transnational corporations on an area is often more harmful than beneficial.

A-grade answer

Transnational corporations are companies operating in more than one country. Usually they have parts of their organisations in several different countries. They can be manufacturing industries such as Toyota and BP but there are also many transnational firms which are involved in services. The major banks operate across the world such as HSBC and Barclays. There are companies which operate in the primary sector as well such as big mining firms such as Rio Tinto Zinc. They have their headquarters based in the country they started in but with the New International Division of Labour (NIDL) they usually have branch plants or subsidiaries in other countries. When they invest money abroad this is known as foreign direct investment (FDI). This money is intended to make more money for the TNC but it can also have positive and negative effects on the area where it is happening.

Toyota started just before the Second World War and after the war developed into the world's biggest car maker. At first it only made cars in Japan but now it operates in every continent. In the 1980s it built factories in the USA and then it has expanded into Europe. There is a Toyota car plant at Burnaston near Derby and another factory near Chester. This has meant that it has invested millions of pounds in the UK as FDI. This has meant advantages and disadvantages for the UK. The advantages are that there are jobs provided by Toyota. This helps reduce unemployment in these two areas as they were places where unemployment rates were high because local industries had shut down. This means that there is a positive multiplier effect as the people with jobs spend money creating jobs for other people and so on. There is also an advantage for the government as the people in work pay tax and do not have unemployment benefit. Toyota also pays tax which gives money to spend on healthcare and schools. Toyota also buys lots of the components for its cars from suppliers in the UK and this creates more jobs and money. The disadvantage of this is that British car makers have not competed with Toyota and have closed down. Toyota may decide to close its British factories if there is a serious recession as TNCs tend to shut their branch plants before closing factories in their home country. Another disadvantage is that most of the profits do not stay in the country where the branch plant is but are taken back to the original country. This is called 'leakage' and is a serious problem with the tourist industry. Many large hotels and tour operators are from developed countries and when they operate in developing countries such as Gambia and Kenya they create some wealth there but most of the money they make is taken back to the developed countries.

Some TNCs pay very low wages when they operate in developing countries and NICs. Making trainers is an industry where there have been problems of exploitation of workers, especially females. The factory making shoes for Fila in Indonesia has

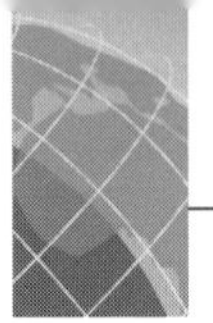

mostly women workers. They have to work very long hours for low wages at much less than a dollar an hour. They have to work overtime and only have one day off a week. Fila is based in South Korea, an NIC, but Indonesia is a developing country. So although there are wages most of the profit is taken back to the HQ. There are also few laws about the environment in Indonesia and so the factory is polluting. This costs Fila less money but affects the health of the workers and other people living next to the factory. Nearly all the managers in the factory are from South Korea and these are the higher paid jobs.

There are many advantages of FDI by TNCs but also some disadvantages. If you are a developing country then it is probably worth having the jobs as the people will not have one if the TNC does not locate there. However, there can be problems if the TNC shuts its branch plants as the government can do very little to stop this happening. There can be effects on the environment as well. Generally it is probably better to have TNCs in an area than having none.

This is a sustained attempt to engage with the topic. The candidate has probably made a brief plan before writing the answer and so the focus is maintained throughout. The introduction is sensible as it defines some terms, TNC and FDI, as well as giving some appropriate examples. The answer then moves into a lengthy section based on Toyota. The material is used to answer directly the question set with some advantages and disadvantages outlined. At the end of this section mention is made of TNCs in the tourism sector. Although relevant, this material should have a separate paragraph and more exemplification. The comments about Fila are well made and again deal directly with the pros and cons of the issue.

The response is well structured with an introduction, a core discussion and a conclusion. In places the grammar could be more secure and more paragraphs would help define the argument as it proceeds. The conclusion deals directly with the question and makes an interesting point about the role of TNCs in the developing world. The three elements of the assessment would all reach Level 3 but not quite at the very top end. The marks would be 8/9 for AO1, 14/17 for AO2 and 4/4 for AO3, giving an overall total of 26/30 — a clear A grade, but not quite an A*.

Question 12

Development and inequalities

How and to what extent is it possible to reduce social and economic inequalities?

A*-grade answer

There are differences in social and economic standards between many different groups of people. The world is divided into the more developed and less developed areas. The 'North' has the highest levels of development in economic and social terms. The high-income group as defined by the World Bank has about 15% of the world's population and 75% of the world's income. At the other end of the scale are the low-income countries, who have 20% of the world's population but only about 1.5% of its income. There are also major contrasts in social factors such as healthcare and education. Access to doctors and schools is much lower in the low-income countries.

Globally there are many organisations trying to reduce inequalities. The United Nations have several units that have development programmes. UNESCO is part of the UN which promotes education and helps teacher training in parts of the less developed world. If children go to school for longer they are more likely to get better-paid jobs. This will raise their standard of living and mean they are more likely to afford better housing, more and better food and suffer less illness. It is especially important for females to be more educated as this can be a way of reducing inequality between males and females, which is a serious issue in some parts of the world. In some countries such as Sudan and Pakistan, women are not always allowed to go to school or university. They have very restricted lives and so it is very difficult to change the culture so that they improve their quality of life.

There are development agencies such as the World Bank which lend money to finance schemes such as road building and irrigation projects. The Three Gorges project in China was partly financed by the World Bank. It produces vast quantities of electricity as well as helping improve navigation along the Yangtze river. This is intended to help China develop as although its economy is growing rapidly it is still a relatively low-income country. However, the development gap between the rich and poor countries has been getting wider in many cases. In particular, sub-Saharan Africa has been struggling to develop. Although many of them are better off than they were a few decades ago, they are still very poor and have low standards of living. One reason is that drought has been very severe in countries such as Mali and Chad. There has been desertification which makes it even more difficult to develop as the productivity of the area reduces. Global warming is likely to make the situation worse in the future. They also have some of the highest rates of population increase with an average of 2.3% increase. The birth rates are about the highest in the world as they are between 30 and 40 per 1,000. This means that even if the economy grows, there are many more people to be supported and so the inequality becomes more not less.

HIV and AIDS is a major issue in many poorer countries although it is also widespread in places such as Russia. The healthcare needed is enormous and costs vast sums of money. It also means that many families do not earn enough money as either the

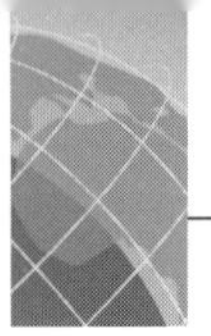

mother or the father or both are ill or have died. One problem has been the cost of the drugs used to treat AIDS but the drug companies are beginning to make them available at cheaper prices so that more people can afford them.

There is an unequal balance in the way trade is carried out around the world. World trade is dominated by TNCs who buy and sell primary and manufactured goods and trade them around the world. There is also a growing trade in services such as banking. The problem is that some countries mainly trade primary products such as iron ore, coffee and bananas. For example Burundi relies upon tea and coffee exports for 90% of its earnings. These have very little added value and so are not worth as much as the products that are made from them. Countries trading manufactured products earn more money. Trade is sometimes limited by tariffs and in some countries the government gives subsidies to an industry which makes its products cheaper. The World Trade Organization tries to free up trade between countries but has struggled to get international agreements which allow the poorer countries to have a better deal. The Fair Trade organisation negotiates higher prices for farmers who join its scheme. Consumers pay a higher price for something that is fair trade but this way the farmers receive a more fair share of the final price. Sainsburys for example only sells fair-trade bananas. This is one way in which inequalities can be reduced but lots of small farmers are not in this scheme and so it only affects a small number of people.

There are many ways in which inequalities can be reduced. Both economic and social inequalities are linked so if one is reduced it often results in the other one reducing. However, it is difficult in some situations to get rid of inequality. The development gap at the global scale is growing despite all the aid that goes to low-income countries, but without it the inequality would be even greater. There is still a long way to go before inequalities are eliminated — if they ever can be.

This is an impressive discussion both in terms of structure and content. It sets the scene at the global scale in the opening paragraph and then goes on to examine methods used to reduce inequalities. The answer deals with social and economic inequalities, which are important if it is to reach Level 3. It is evaluative in its style, noting the measures taken and then offering some assessment of their effectiveness. The answer keeps a sharp focus on the question and offers a wide-ranging selection of material. The conclusion links back to the question. This style of response would receive full marks in all three sections of the mark scheme, and would therefore gain an A* grade.